LIGNOCELLULOSIC PRECURSORS USED IN THE SYNTHESIS OF ACTIVATED CARBON - CHARACTERIZATION TECHNIQUES AND APPLICATIONS IN THE WASTEWATER TREATMENT

Edited by **Virginia Hernández Montoya and Adrián Bonilla Petriciolet**

Lignocellulosic Precursors Used in the Synthesis of Activated Carbon - Characterization Techniques and Applications in the Wastewater Treatment

Edited by Virginia Hernández Montoya and Adrián Bonilla Petriciolet

Contributors

A. Alicia Peláez-Cid, M.M. Margarita Teutli-León, Virginia Hernández-Montoya, Josafat García-Servin, José Iván Bueno-López, Carlos J. Durán-Valle, María del Rosario Moreno-Virgen, Rigoberto Tovar-Gómez, Didilia I. Mendoza-Castillo, Adrián Bonilla-Petriciolet, Rosa Miranda, César Sosa, Diana Bustos, Eileen Carrillo and María Rodríguez-Cantú

Published by InTech

Janeza Trdine 9, 51000 Rijeka, Croatia

Publishing Process Manager Sandra Bakic
Typesetting InTech DTP Team
Cover InTech Design Team

Additional hard copies can be obtained from orders@intechopen.com

Lignocellulosic Precursors Used in the Synthesis of Activated Carbon -
Characterization Techniques and Applications in the Wastewater Treatment,
Edited by Virginia Hernández Montoya and Adrián Bonilla Petriciolet
p. cm.
ISBN 978-953-51-0197-0

For EU product safety concerns:
IN TECH d.o.o., Prolaz Marije Krucifikse Kozulić 3, 51000 Rijeka, Croatia,
info@intechopen.com or visit our website at intechopen.com.

Meet the editors

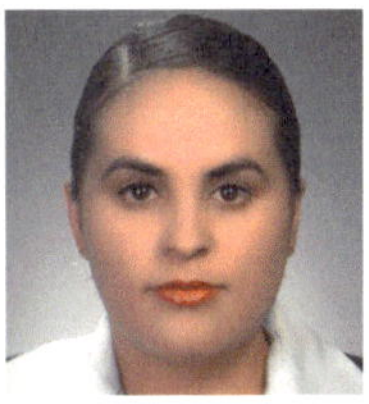

Virginia Hernández Montoya completed her Ph.D. at the Benemérita Universidad Autónoma de Puebla, México. Her research interests include the preparation and characterization of carbonaceous adsorbents and their application in the removal of pollutants from water and air. Dr. Hernández-Montoya has published several papers in International Journals and she has participated in International Conferences, Symposiums and Congress. Actually, Dr. Hernández-Montoya is Professor in the Department of Chemical Engineering of Instituto Tecnológico de Aguascalientes (México) and she collaborates with national and international research groups.

Adrián Bonilla-Petriciolet is Professor in the Department of Chemical Engineering of Instituto Tecnológico de Aguascalientes, México. His research interests are applied thermodynamics, process modeling, materials science, and optimization. He has published several papers in international journals and presented a number of papers in conferences.

Contents

Preface

The synthesis and characterization of activated carbons (ACs) obtained from lignocellulosic precursors is a topic widely studied by a number of researchers worldwide. In the last decades, an increase has been observed in the number of publications related to the synthesis, modification, characterization and application of ACs obtained from lignocellulosic materials. Particularly, the applications of these carbons are primarily focused in the removal of several inorganic and organic pollutants from water and wastewaters.

In this context, the purpose of this book is to provide, for interested readers in the topic of activated carbons, the actual or alternative lignocellulosic precursors used in the elaboration of ACs (shells, stones, seeds, woods, etc.), the different methods and experimental conditions employed in their synthesis; the recent and more specialized techniques used in the characterization of ACs and the specific physicochemical characteristics that activated carbon must show to remove efficiently priority pollutants from water. Also, the importance of pyrolysis method for energy and carbon production is discussed in this book.

The book contains five chapters and a short description is given in the following points:

- Chapter 1: Provides a twenty-year (1992 – 2011) worldwide research review regarding a large amount of lignocellulosic materials proposed as potential precursors in the production of activated carbon.

- Chapter 2: Describes the principal methods used in the preparation of activated carbons from lignocellulosic materials by chemical and physical procedures. An analysis of the experimental conditions used in the synthesis of ACs has been made attending to the carbon specific surface area. Also, the advantages and disadvantages of each method are discussed.

- Chapter 3: Introduces the basic principles of the common techniques used in the characterization of activated carbons. For example, this chapter includes techniques to determine textural parameters such as mercury porosimetry and gas adsorption isotherms; and different spectroscopies to determine chemical functionality (Raman, FT-IR, etc.) and other X-Ray techniques.

- Chapter 4: Provides an overview of the application of activated carbons obtained from lignocellulosic precursors for wastewater treatment. Analysis and discussion are focused on the performance of different activated carbons obtained from several precursors and their advantages and capabilities for the removal of relevant toxic compounds and pollutants from water.

- Chapter 5: Analyses the use of pyrolysis for the valorization of two Mexican typical agricultural wastes (orange peel and pecan nut shell) for energy and carbon production. Also, the analysis of pyrolysis yields for various biomasses at different conditions is reported and, finally the composition of the liquid fractions (i.e., bio-oil) obtained from the pyrolysis of orange peel and pecan nut shell were analysed.

I would like to thank all the authors for their excellent contributions to this book and to Instituto Tecnológico de Aguascalientes for the facilities to work in this project.

Ph.D Virginia Hernández Montoya
Instituto Tecnológico de Aguascalientes
México

1

Lignocellulosic Precursors Used in the Elaboration of Activated Carbon

A. Alicia Peláez-Cid and M.M. Margarita Teutli-León
Benemérita Universidad Autónoma de Puebla
México

1. Introduction

Many authors have defined activated carbon taking into account its most outstanding properties and characteristics. In this chapter, activated carbon will be defined stating that it is an excellent adsorbent which is produced in such a way that it exhibits high specific surface area and porosity. These characteristics, along with the surface's chemical nature (which depends on the raw materials and the activation used in its preparation process), allow it to attract and retain certain compounds in a preferential way, either in liquid or gaseous phase. Activated carbon is one of the most commonly used adsorbents in the removal process of industrial pollutants, organic compounds, heavy metals, herbicides, and dyes, among many others toxic and hazardous compounds.

The world's activated carbon production and consumption in the year 2000 was estimated to be 4 x 10^8 kg (Marsh, 2001). By 2005, it had doubled (Elizalde-González, 2006) with a production yield of 40%. In the industry, activated carbon is prepared by means of oxidative pyrolysis starting off soft and hardwoods, peat, lignite, mineral carbon, bones, coconut shell, and wastes of vegetable origin (Girgis et al., 2002; Marsh, 2001).

There are two types of carbon activation procedures: Physical (also known as thermal) and chemical. During physical activation, the lignocellulosic material as such or the previously carbonized materials can undergo gasification with water vapor, carbon dioxide, or the same combustion gases produced during the carbonization. Ammonium persulfate, nitric acid, and hydrogen peroxide have also been used as oxidizing agents (Salame & Bandoz, 2001). Chemical activation consists of impregnating the lignocellulosic or carbonaceous raw materials with chemicals such as $ZnCl_2$, H_3PO_4, HNO_3, H_2SO_4, NaOH, or KOH (Elizalde-González & Hernández-Montoya, 2007; Girgis et al., 2002). Then, they are carbonized (a process now called "pyrolysis") and, finally, washed to eliminate the activating agent. The application of a gaseous stream such as air, nitrogen, or argon is a common practice during pyrolysis which generates a better development of the material's porosity. Although not commonly, compounds such as potassium carbonate, a cleaner chemical agent (Tsai et al., 2001b; 2001c) or formamide (Cossarutto et al., 2001) have been also used as activating agents.

Commercial activated carbon is produced as powder (PAC), fibers (FAC), or granules (GAC) depending on its application. It regularly exhibits BET specific surface magnitudes between 500 and 2000 m^2g^{-1}. However, the so-called "super-activated carbons" exhibit surfaces areas above 3000 m^2g^{-1}. Activated carbon's macro, meso, and micropore volumes may range from 0.5 to 2.5 cm^3g^{-1} (Marsh, 2001).

The adsorption capabilities of activated carbon are very high because of its high specific surface, originated by porosity. Also, depending on what type of activation was used, the carbon's surface may exhibit numerous functional groups, which favor the specific interactions that allow it to act as an ionic interchanger with the different kinds of pollutants.

The activated carbon is commonly considered an expensive material because of the chemical and physical treatments used in its synthesis, its low yield, its production's high energy consumption, or the thermal treatments used for its regeneration and the losses generated meanwhile. However, if its high removal capacity compared to other adsorbents is considered, the cost of production does not turn out to be very high. The search for the appropriate mechanism for its pyrolysis process is an important factor for tackling production costs.

The exhausted material's thermal regeneration (Robinson et al., 2001) consists of drying the wet carbon, pyrolysis of the adsorbed organic compounds, and reactivating the carbon, which generates mass losses up to 15 %. The carbon's regeneration can also be accomplished by using water vapor or solvents to desorb the absorbed substances, which, in turn, leads to a new problem regarding pollution. Because of these environmental inconveniences as well as the loss in adsorption capacity and the increase in costs which the regeneration process implies, using new carbon once the old one's surface has been saturated is often preferred.

With the goal of diminishing the cost of producing activated carbon, contemporary research is taking a turn towards industrial or vegetable (lignocellulosic) wastes to be used as raw material, and, then, lessen the cost of production (Konstantinou & Pashalidis, 2010). Besides, the use of these precursors reduces residue generation in both rural and urban areas.

This chapter presents a twenty-year (1992 – 2011) worldwide research review regarding a large amount of lignocellulosic materials proposed as potential precursors in the production of activated carbon. The most common characteristics that lignocellulosic wastes used in carbon production and the parameters that control porosity development and, hence, the increase in specific surface during carbonization are also mentioned. A comparison between countries whose scientists are interested in carbon preparation from alternative waste lignocellulosic materials by continent is made. The most commonly used agents for chemical, physical, or a combination of both activations methods which precursors undergo are shown.

2. Characteristics of the selected raw materials for activated carbon production

The materials selected nowadays to be potential precursors of activated carbons must fulfill the following demands:

1) They must be materials with high carbon contents and low inorganic compound levels (Tsai et al., 1998) in order to obtain a better yield during the carbonization processes. This is valid for practically every lignocellulosic wastes. They must be plentiful in the region or country where they will be used to solve any specific environmental issue. For example, corncob has been used to produce activated carbon and, according to Tsai et al. (1997), corn grain is a very important agricultural product in Taiwan. The same condition applies for the avocado, mango, orange, and guava seeds in Mexico (Elizalde-González et al., 2007; Elizalde-González & Hernández-Montoya, 2007, 2008, 2009a, 2009b, 2009c; Dávila-Jiménez et al., 2009). Specifically, Mexico has ranked number one in the world for avocado production, number two for mango, and number four for orange (Salunkhe & Kadam, 1995). On the other hand, jute stick is abundantly available in Bangladesh and India (Asadullah et al., 2007), from which bio-oil is obtained, and the process's residue has been used to produce activated carbon. Bamboo, an abundant and inexpensive natural resource in Malaysia, was also used to prepare activated carbon (Hameed et al., 2007). Cherry pits are an industrial byproduct abundantly generated in the Jerte valley at Spain's Caceres province (Olivares-Marín et al., 2006). Other important wastes generated in Spain that have also been proposed with satisfying results in the production of activated carbon with high porosity and specific surface area are: olive-mill waste generated in large amounts during the manufacture of olive oil (Moreno-Castilla et al., 2001) and olive-tree wood generated during the trimming process of olive trees done to make their development adequate (Ould-Idriss et al., 2011).

2) The residue generated during consumption or industrial use of lignocellulosic materials regularly represents a high percentage of the source from which it is obtained. For example, mango seed is around 15 to 20 % of manila mango from which it is obtained (Salunkhe & Kadam, 1995). In the case of avocado, 10 to 13 % of the fruit weight corresponds to the kernel seed and it is garbage after consumption (Elizalde-González et al., 2007). Corn cob is approximately 18 % of corn grain (Tsai et al., 2001b). Orange seeds constitute only about 0.3 % of the fresh mature fruit (Elizalde-González & Hernández-Montoya, 2009c), but orange is the most produced and most consumed fruit worldwide (Salunkhe & Kadam, 1995). Sawdust does not constitute a net percentage of tree residue, rather, it is a waste obtained from wood applications conditioning. However, it has proven to be a good precursor when it is obtained from mahogany (Malik, 2003).

3) They must be an effective and economic material to be used as an adsorbent for the removal of pollutants from both gaseous and liquid systems. Specifically, carbons produced from lignocellulosic precursors have been used to eliminate basic dyes (Elizalde-González et al., 2007; Elizalde-González & Hernández-Montoya, 2007; Girgis et al., 2002; Hameed et al., 2007; Rajeshwarisivaraj et al., 2001), acid dyes (Elizalde-González et al., 2007; Elizalde-González & Hernández-Montoya, 2008, 2009a, 2009b, 2009c; Malik, 2003; Rajeshwarisivaraj et al., 2001; Tsai et al., 2001a), reactive dyes (Elizalde-González et al., 2007; Senthilkumaara et al., 2006), direct dyes (Kamal, 2009; Namasivayam & Kavitha, 2002; Rajeshwarisivaraj et al., 2001), metallic ions such as Cr^{4+}, Hg^{2+} and Fe^{2+} (Rajeshwarisivaraj et al., 2001), Eu^{3+}

(Konstantinou & Pashalidis, 2010), Cu^{2+} (Dastgheib & Rockstraw, 2001; Konstantinou & Pashalidis, 2010; Toles et al., 1997) or Pb^{2+} (Giraldo & Moreno-Piraján, 2008), and low molecular mass organic compounds such as phenol (Giraldo & Moreno-Piraján, 2007; Wu et al., 1999, 2001), chlorophenol (Wu et al., 2001), and nitro phenol (Giraldo & Moreno-Piraján, 2008). For example, bamboo powder charcoal has demonstrated being an attractive option for treatment of superficial and subterranean water polluted by nitrate-nitrogen (Mizuta et al., 2004). Carbon produced from bamboo waste (Ahmad & Hammed, 2010) as well as the one obtained from avocado peel (Singh & Kumar, 2008) have proven effective in diminishing COD during the treatment of cotton textile mill wastewater and wastewater from coffee processing plant, respectively. Carbon molecular sieves for separating gaseous mixtures are another application of activated carbons prepared from lignocellulosic precursors (Ahmad et al., 2007; Bello et al., 2002).

3. Parameters for activated carbon preparation

Research has shown that carbons's properties such as specific surface area, porosity, density and mechanical resistance depend greatly on the raw material used. However, it may be possible to modify these parameters changing the conditions in the pyrolysis process of the lignocellulosic materials.

In particular, the most important parameters to be considered while preparing activated carbons from lignocellulosic materials are described below.

3.1 Activating agent

H_3PO_4 is the most commonly used chemical agent for synthesis of activated carbon. The use of $ZnCl_2$ has declined because of the environmental pollution problems with zinc disposal (Girgis et al., 2002). In the case of physical activation, the use of water vapor and carbon dioxide is preferred to promote the partial oxidation of the surface instead of oxygen, which is too reactive.

3.2 Mass ratio of precursor and activating agent

The complete saturation of lignocellulosic precursor must be ensured to develop the adsorbent porosity with the minimum activating agent consumption. This leads a minor consumption of chemical compounds and a better elimination of the excess during the carbon washing process. The effect of the increase in proportion of the impregnation over the carbon porous structure is greater than the one obtained with the increase of carbonizing temperature (Olivares-Marín et al., 2006a).

3.3 Heating speed

Regularly, heating ramps with a low speed are used for preparation of activated carbon. This approach allows the complete combustion of material precursor and favors a better porosity development. Rapid heating during pyrolysis produces macroporous residue (Heschel & Klose, 1995).

3.4 Carbonizing temperature

It has the most influence over the activated carbon's quality during the activating process. It must be at least 400 °C to ensure the complete transformation of organic compounds (present in lignocellulosic precursors) into graphene structures. The degree of specific surface area development and porosity is incremented on par with the carbonizing temperature (Olivares-Marín et al., 2006b). During physical activation, carbonization temperatures are greater than those needed for chemical activation (Lussier et al., 1994). However, carbonization temperatures used in activated carbon production are generally greater than 400 °C and temperatures ranging from 120 to 1000 °C have been used. (Elizalde et al., 2007; Elizalde-González & Hernández-Montoya, 2008; Rajeshwarisivaraj et al., 2001; Salame & Bandosz, 2001). It has been reported that carbon obtained from peach pits with temperatures below 700 °C still have a high content of hydrogen and oxygen (MacDonald & Quinn, 1996).

3.5 Carbonizing time

This parameter must be optimized to obtain the maximum porosity development while still minimizing the material's loss due to an excessive combustion. Bouchelta et al. (2008) have shown that the yield percentage decreases with increase of activation temperature and hold time. Carbonization times ranging from 1 h (Rajeshwarisivaraj et al., 2001; Wu et al., 1999) up to 14 h (Rajeshwarisivaraj et al., 2001) have been used in charcoal production.

3.6 Gas flow speed

It has been observed that during pyrolysis, the passing on an inert gas, such as N_2 or Ar, favors the development in the carbon's porosity. In this case, the flow and the gas type may affect the final properties of the activated carbon. CO_2 flow-rate had a significant influence on the development of the surface area of oil palm stones (Lua & Guo, 2000).

3.7 Effect of washing process

During the lignocellulosic residue's pyrolysis, the presence of chemical activating agents generates carbons with a more orderly structure. The later elimination of chemical activating agents, by means of successive washings, will allow a better development of porosity.

4. Worldwide studied precursors

Numerous lignocellulosic residues have been selected as potential activated carbon precursors. Among them, there is the wood obtained from several kinds of tree species such as *Eucalyptus* (Bello et al., 2002; Ngernyen et al., 2006; Rodrígez-Mirasol et al., 1993), pine (Giraldo & Moreno-Piraján, 2007; Sun et al., 2008), *Quercus agrifolia* (Robau-Sánchez et al., 2001), wattle (Ngernyen et al., 2006), china fir (Zuo et al., 2010), acacia (Kumar et al., 1992), olive tree (Ould-Idriss et al., 2011), softwood bark (Cao et al., 2002), mahogany sawdust (Malik, 2003), sawdust flash ash (Aworn et al., 2008), and sawdust (Giraldo & Moreno-Piraján, 2008; Zhang et al., 2010), coconut shell (Cossarutto et al., 2001; Giraldo & Moreno-Piraján, 2007; Hayashi et al., 2002; Heschel & Klose, 1995; Hu et al., 2001; Kannan & Sundaram, 2001), coconut fiber (Namasivayam & Kavitha, 2002; Phan et al., 2006; Senthilkumaara et al., 2006), corn cob (Aworn et al., 2008; Tsai et al., 1997; 1998; 2001a;

2001b; Tseng & Tseng, 2005; Wu et al., 2001), cherry stones (Gergova et al., 1993; 1994; Heschel & Klose, 1995; Lussier et al., 1994; Olivares-Marín et al., 2006a; 2006b), apricot stones (Gergova et al., 1993; 1994), peach stones (Heschel & Klose, 1995; MacDonald & Quinn, 1996; Molina-Sabio et al., 1995; 1996; Rodríguez-Reinoso & Molina-Sabio, 1992) and peach seed (Giraldo & Moreno-Piraján, 2007), mixture of apricot and peach stones (Puziy et al., 2005), wheat straw (Kannan & Sundaram, 2001), rice straw (Ahmedna et al., 2000) and rice husks (Ahmedna et al., 2000; Aworn et al., 2008; Kalderis et al., 2008; Kannan & Sundaram, 2001; Malik, 2003; Swarnalatha et al., 2009), sugarcane bagasse (Ahmedna et al., 2000; Aworn et al., 2008; Giraldo & Moreno-Piraján, 2007; Juang et al., 2002; 2008; Kalderis et al., 2008; Tsai et al., 2001;), palm fiber (Guo et al., 2008), palm pit (Giraldo & Moreno-Piraján, 2007; 2008), palm shell (Ahmad et al., 2007; Arami-Niya et al., 2010; Hayashi et al., 2002), stem of date palm (Jibril et al., 2008), and palm seeds (Gou et al., 2008; Hu et al., 2001), palm stones (Lua & Guo, 2000), pecan shells (Ahmedna et al., 2000; Dastgheib & Rockstraw, 2001; Toles et al., 1997), almond shells (Gergova et al., 1994; Hayashi et al., 2002; Iniesta et al., 2001; Mourao et al., 2011; Nabais et al., 2011; Rodríguez-Reinoso & Molina-Sabio, 1992; Toles et al., 1997), macadamia shells (Aworn et al., 2008; Evans et al., 1999), cedar nut shells (Baklanova et al., 2003), hazelnut shells (Heschel & Klose, 1995), pistachio shell (Hayashi et al., 2002), and walnut shells (Hayashi et al., 2002; Heschel & Klose, 1995), bamboo powder (Ahmad & Hameed, 2010; Hammed et al., 2007; Kannan & Sundaram, 2001; Mizuta et al., 2004), jute fibers (Asadullah et al., 2007; Phan et al., 2006; Senthilkumaara et al., 2006), plum kernels (Heschel & Klose, 1995; Wu et al., 1999), avocado kernel seeds (Elizalde-González et al., 2007) and avocado peel (Devi et al., 2008), coffee bean husks (Baquero et al., 2003), coffee residue (Boudrahem et al., 2009), and coffee ground (Evans et al., 1999), date stones (Bouchelta et al., 2008; Hazourli et al., 2009), grape seeds (Gergova et al., 1993, 1994), vine shoot (Mourao et al., 2011), orange seeds (Elizalde-González & Hernández-Montoya, 2008, 2009c) and guava seeds (Elizalde-González & Hernández-Montoya, 2008, 2009a, 2009b), mango pit (husk and seed) (Dávila-Jimenez et al., 2009; Elizalde-González & Hernández-Montoya, 2007; 2008), olive stones (Rodríguez-Reinoso & Molina-Sabio, 1992; Yavuz et al., 2010) and olive cake (Konstantinou & Pashalidis, 2010; Moreno-Castilla et al., 2001), peanut hull (Girgis et al., 2002; Kannan & Sundaram, 2001), cassava peel (Rajeshwarisivaraj et al., 2001), pomegranate peel (Amin, 2009), cotton stalks (Girgis & Ishak, 1999), kenaf (Valente-Nabais et al., 2009), cork waste (Carvalho et al., 2004), flamboyant pods (A.M.M. Vargas et al., 2011), rapeseed (Valente-Nabais et al., 2009), *Macuna musitana* (Vargas et al., 2010), and seed husks of *Moringa Oleifera* (Warhurst et al., 1997). Table 1 shows clearly the lignocellulosic precursors used in activated carbon production classified according to the source they were obtained from.

Figure 1 shows the great variety of lignocellulosic residues used in worldwide production of activated carbon. It can be observed that wood from several tree species, several kinds of nuts, or different coconut parts are among the most commonly used along with the traditional raw materials used for the preparation of activated carbon. This figure shows that from a single vegetable, different parts have been tested as precursors. For example, the seed and peel of avocado have been studied (Elizalde et al., 2007; Singh & Kumar, 2008). The same condition applies for the rice straw (Ahmedna et al., 2000) and the rice husk (Kalderis et al., 2008; Swarnalatha et al., 2009). Note that when carbons are prepared with lignocellulosic precursors, they are called charcoal. If they are of mineral origin, then they are called coal. Both kinds are susceptible to chemical, physical, or a combination of both activation types to produce the outstanding activated carbons.

It has been found that the activated carbon's properties depend greatly on the composition of their raw materials (Gergova et al., 1993; Girgis et al., 2002). Development of porosity and active sites with a specific character is aided by physical activation because a partial oxidation occurs, and the carbon's surface is enriched with several functional groups (Salame & Bandoz, 2001). Chemical activation further develops these characteristics. Additionally, chemical activation has several advantages over physical activation. Besides, it is done at lower temperatures. Some authors have chosen a combination of both methods to produce their activated carbons for fitting specific applications. For example, it can be cited the activated carbon obtained from coconut peel activated with water vapor and then treated with formamide to accomplish the adsorption of the vapor (Cossarutto et al., 2001). On the other hand, there are wood carbons chemically activated with H_3PO_4 and KOH, and then treated with ammonia persulfate, nitric acid, or hydrogen peroxide (as oxidating agents) with the objective of obtaining carbons either with the nitro- group with positive charges on the nitrogen atom or with negative charges on the oxygen atoms, making them better adsorbents for ionic species (Salame & Bandoz, 2001).

Wood	*Eucalyptus*	Nuts Shells	Pecan	Palm	Fiber
	Pine		Almond		Pit
	Quercus agrifolia		Macadamia		Shell
	Wattle		Cedar		Stem of date
	China fir		Hazelnut		Seeds
	Acacia		Pistachio		Stones
	Olive tree		Walnut	Coconut	Shell
	Softwood bark	Stones	Cherry		Fiber
	Mahogany sawdust		Apricot	Straw	Rice
	Sawdust flash ash		Peach		Wheat
	Sawdust		Plum	Peel	Avocado
	Peach		Date		Cassava
	Plum		Olive		Pomegranate
Seeds	Avocado	Husks	Rice	Jute	Fibers
	Grape		Coffee bean		Stick
	Orange		Mango	Coffee	Ground
	Guava		*Moringa Oleifera*		Residue
	Mango	Corncob	Peanut hull	Rapeseed	Cork waste
	Macuna musitana	Kenaf	Cotton stalks	Flamboyant pods	
	Sugarcane bagasse	Vine shoot	Olive cake	Bamboo powder	

Table 1. Waste materials used in activated carbon production grouped according to their source.

Although some carbons obtained from corn cob with a BET specific surface up to 2595 m^2g^{-1} have been prepared via chemical activation with KOH (Tseng & Tseng, 2005), high surface areas can be obtained by means of physical activation. These carbons reach values of 1400 m^2g^{-1} or more using *Eucaliptus* as the precursor and CO_2 as an oxydating agent (Ngernyen et al., 2006; Rodríguez-Mirasol et al., 1993). Figure 2 shows that the worldwide tendency in relationship with the activation type indicates that activated carbons are physically prepared in greater amounts. This tendency may be due to the fact that the best activated carbons for adsorbing of species with positive charges are those oxidized with acid functional groups. The development of these acid groups can be done via oxidation with oxygen present in the air or using some other oxidating materials such as water vapor or carbon dioxide (Dastgheib & Rockstraw, 2001). Besides, with physical activation, there is no consumption of chemical activating agents. This simplifies the preparation of activated carbons in terms of avoiding the washing procedure involved in the chemical activation and the pollution caused by this procedure.

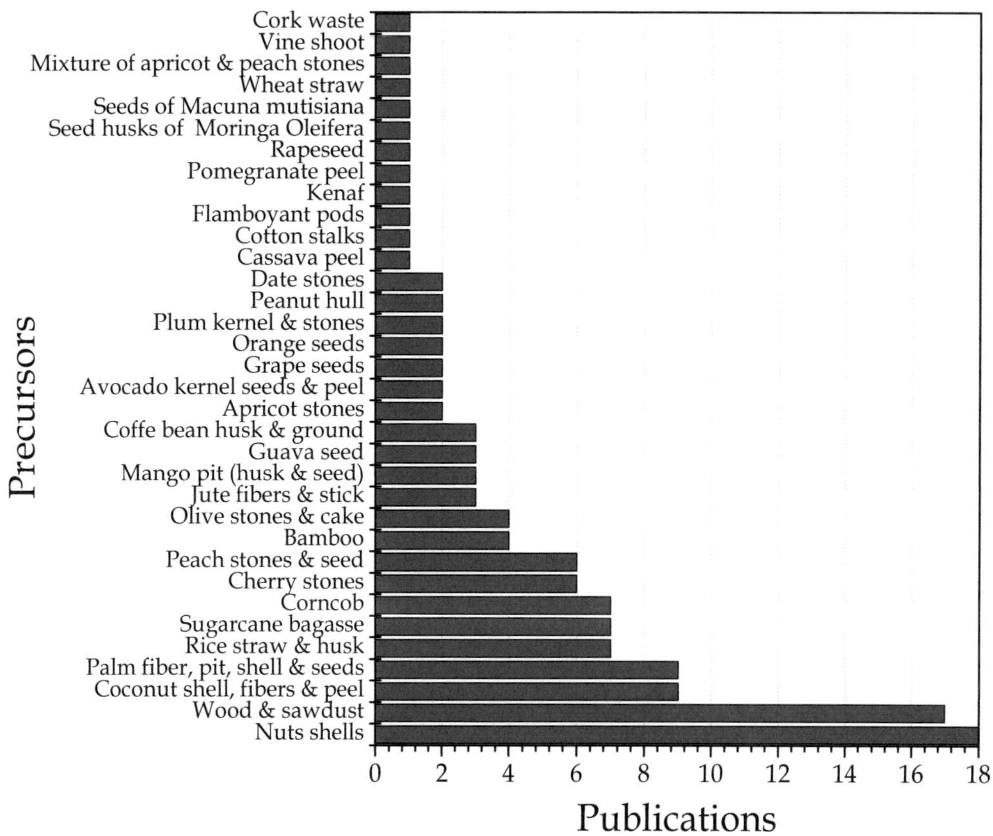

Figure 1. Lignocellulosic raw materials used in the production of activated carbon. Wood includes several varieties such as Acacia, *Eucalyptus*, fir, mahogany, olive, pine, and wattle. Almond, cedar, hazelnut, macadamia, pecan, pistachio, and, walnut are included in the nuts shells class.

Figure 2 also shows that some authors have also opted for combining activation methods. They use some of the most common chemical agents and then employ streams of diverse oxidating agents in place of inert gases.

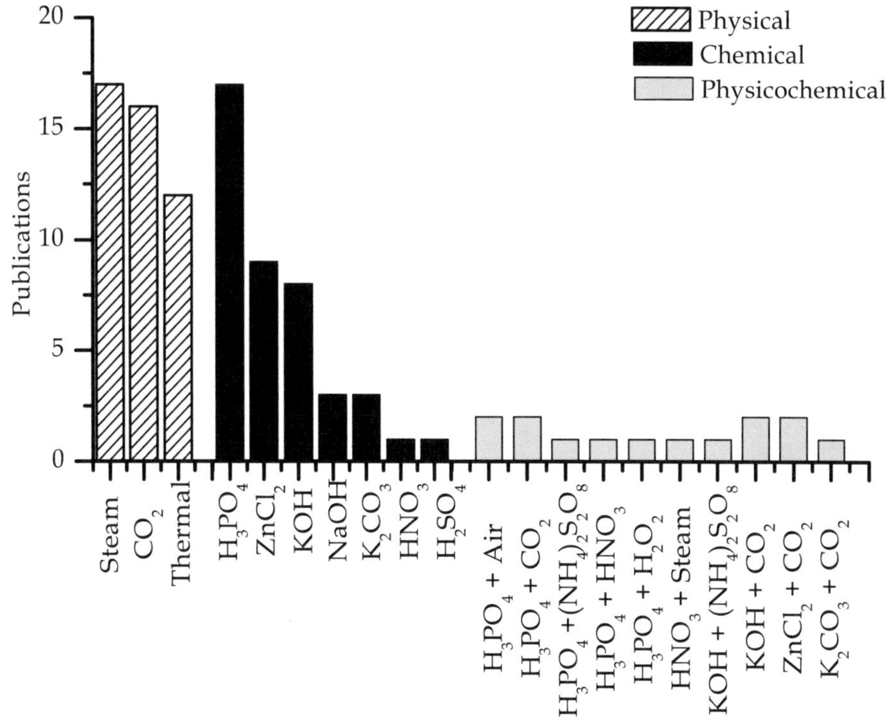

Figure 2. Comparison between the different types of activation and activating agents used in the preparation of activated carbons from lignocellulosic residues.

As a result of the review done, see Figure 3, the different countries' participation in the production of activated carbon was established for this chapter. Asia is the continent with the most research done for the reduction of costs in the production of activated carbon, followed by Europe and America. In Asia, with the exception of Japan, all the countries that participated in the research can be considered underdeveloped, same as America, with the exception of the USA and Canada. It could be thought that the USA has a high degree of research because it is a leading country in terms of technological development in many areas of knowledge. Regarding Europe, it is clear its low participation in this research field. Only Spanish researchers seem to be interested in the activated carbon production problem and they have reported the use of the diverse residues generated in their country for activated carbon preparation. In Africa, because of its underdeveloped economies, only Egypt, Algeria and Moroco participate in this research topic.

Even though the generalized tendency regarding the production of activated carbon leads towards the use of lignocellulosic materials, these can be produced from any carbon-based material (Girgis et al., 2002). Other non-conventional materials that have also been tested are the following: waste slurry of fertilizer plants and blast furnace waste (Gupta et al., 1997), bituminous coal (H. Teng et al., 1997, 1998), paper mill sludge (Khalili et al., 2000), bagasse fly ash (Gupta et al., 2000), waste tires (H. Teng et al., 2000), anthracite (Lillo-Ródenas et al., 2001; Lozano-Castelló et al., 2001), sewage sludge plus coconut husk (Graham et al., 2001;

Tay et al., 2001), sewage sludge (Graham et al., 2001), sewage sludge plus peanut shell (Graham et al., 2001), sewage sludge of derived fertilizer (Bagreev et al., 2001), viscose rayon (Ko et al., 2002), corrugated paper plus silica (Okada et al., 2005), resorcinol-formaldehyde resin (Elsayed et al., 2007), cattle manure compost (Kian et al., 2008), among others.

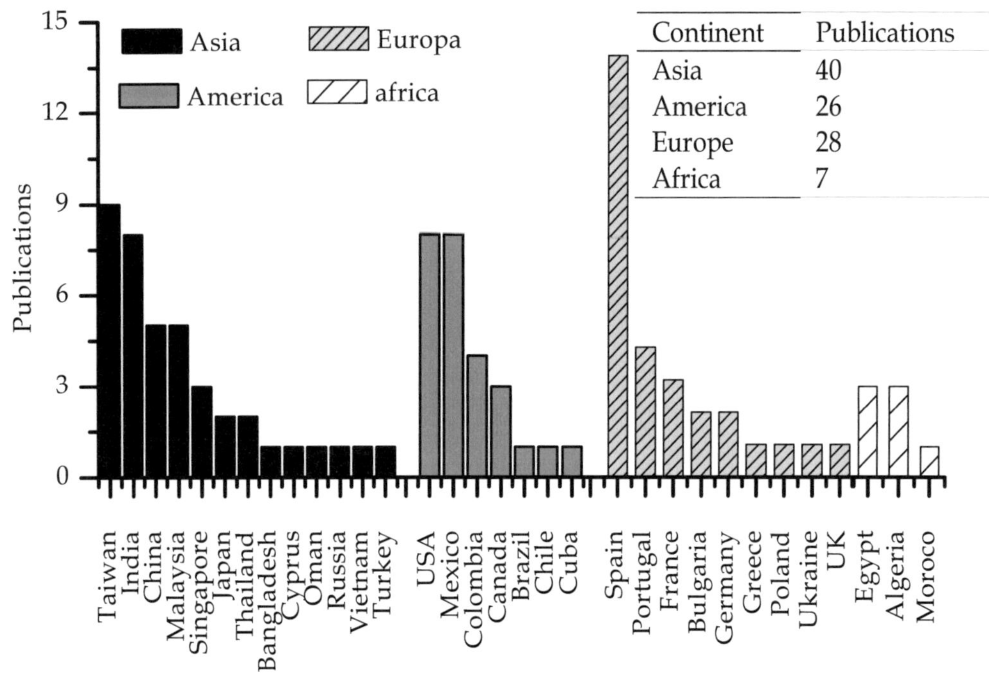

Figure 3. Worldwide distribution and production of activated carbon obtained from lignocellulosic wastes.

5. Conclusion

The literature review (1992 - 2011) indicates that worldwide researchers try to propose new sources to obtain raw materials for the production of activated carbon. They have in mind not only to lessen its cost of production, but also to diminish environmental impact of agricultural and industrial wastes. The way to enhance the adsorptive qualities of the carbons produced is also being studied to make its production more profitable, and, hence, solve specific environmental issues.

6. References

[1] Ahmad, A.A. & Hameed, B.H. (2010). Effect of preparation conditions of activated carbon from bamboo waste for real textile wastewater. *Journal of Hazardous Materials*, Vol. 173, No. 1-3, (January 2010), pp. (487–493), ISSN 0304-3894.

[2] Ahmad, M.A., Wan-Daud, W.M.A. & Aroua, M.K. (2007). Synthesis of carbon molecular sieves from palm shell by carbon vapor deposition. *Journal of Porous Mater*, Vol. 14, No. 4, (March 2007), pp. (393-399), ISSN 0165-2370.

[3] Ahmedna, M., Marshall, W.E. & Rao, R.M. (2000). Production of granular activated carbons from select agricultural by-products and evaluation of their physical, chemical and adsorption properties. *Bioresource Technology*, Vol. 71, No. 2, (January 2000), pp. (113–123) ISSN 0960-8524.

[4] Amin, N.K. (2009). Removal of direct blue-106 dye from aqueous solution using new activated carbons developed from pomegranate peel: Adsorption equilibrium and kinetics. *Journal of Hazardous Materials*, Vol. 165, No. 1-3, (June 2009), pp. (52–62), ISSN 0304-3894.

[5] Arami-Niya, A., Daud, W.M.A.W. & Mjalli, F.S. (2010). Using granular activated carbon prepared from oil palm shell by ZnCl2 and physical activation for methane adsorption. *Journal of Analytical and Applied Pyrolysis*, Vol. 89, No. 2, (November 2010), pp. (197–203), ISSN 0165-2370.

[6] Asadullah, M., Rahman, M.A., Motin, M.A. & Sultan, M.B. (2007). Adsorption studies on activated carbon derived from steam activation of jute stick char. *Journal of Surface Science & Technology*, Vol. 23, No. 1-2, pp. (73–80), ISSN 0970-1893.

[7] Aworn, A., Thiravetyan, P. & Nakbanpote W. (2008). Preparation and characteristics of agricultural waste activated carbon by physical activation having micro- and mesopores. *Journal of Analytical and Applied Pyrolysis*, Vol. 82, No. 2, (July 2008), pp. (279–285), ISSN 0165-2370.

[8] Bagreev, A., Bandosz, T. J. & Locke, D.L. (2001). Pore structure and surface chemistry of adsorbents obtained by pyrolysis of sewage sludge-derived fertilizer. *Carbon*, Vol. 39, No. 13, (November 2001), pp. (1971–1979), ISSN 0008-6223.

[9] Baklanova, O.N., Plaksin, G.V., Drozdov, V.A., Duplyakin, V.K., Chesnokov, N.V., Kuznetsov, B.N. (2003). Preparation of microporous sorbents from cedar nutshells and hydrolytic lignin. *Carbon*, Vol. 41, No. 9, (June 2003), pp. (1793–1800), ISSN 0008-6223.

[10] Baquero, M.C., Giraldo, L., Moreno, J.C., Suárez-García, F., Martínez-Alonso, A. & Tascón, J.M.D. (2003), Activated Carbons by pyrolysis of coffee bean husks in presence of phosphoric acid. *Analytical Applied Pyrolysis*, Vol. 70, No. 2, (December 2003) pp. (779–784), ISSN 0165-2370.

[11] Bello, G., García, R., Arriagada, R., Sepúlveda-Escribano, A., Rodríguez-Reinoso, F. (2002). Carbon molecular sieves from Eucalyptus globulus charcoal. *Microporous and Mesoporous Materials*, Vol. 56, No. 2, (November 2002), pp. (139–145), ISSN 1387-1811.

[12] Bouchelta, C., Medjram, M.S., Bertrand, O. & Bellat, J.P. (2008). Preparation and characterization of activated carbon from date stones by physical activation with steam. *Journal of Analytical and Applied Pyrolysis*, Vol. 82, No. 1, (July 2008), pp. (70–77), ISSN 0165-2370.

[13] Boudrahem, F., Aissani-Benissad, F. & Aït-Amar, H. (2009). Batch sorption dynamics and equilibrium for the removal of lead ions from aqueous phase using activated carbon developed from coffee residue activated with zinc chloride. *Journal of Environmental Management*, Vol. 90, No. 10, (July 2009), pp. (3031–3039), ISSN 0301-4797.

[14] Cao, N., Darmstadt, H., Soutric, F. & Roy, Ch. (2002). Thermogravimetric study on the steam activation of charcoals obtained by vacuum and atmospheric pyrolysis of softwood bark residues. *Carbon*, Vol. 40, No. 4, (April 2002), pp. (471–479), ISSN 0008-6223.

[15] Carvalho, A.P., Gomes, M., Mestre, A.S., Pires, J. & Brotas de Carvalho, M. (2004). Activated carbons from cork waste by chemical activation with K2CO3. Application to adsorption of natural gas components. *Carbon*, Vol. 42, No. 3, (January 2004), pp. (667–69), ISSN 0008-6223.

[16] Cossarutto, L., Zimny, T., Kaczmarczyk, J., Siemieniewska, T., Bimer, J., & Weber, J.V. (2001). Transport and sorption of water vapour in activated carbons. *Carbon,* Vol. 39, No. 15, (December 2001), pp. (2339–2346), ISSN 0008-6232.
[17] Dastgheib, S.A. & Rockstraw, D.A. (2001). Pecan shell activated carbon: synthesis, characterization, and application for the removal of copper from aqueous solution. *Carbon*, Vol. 39, No. 12, (October 2001), pp. (1849–1855), ISSN 0008-6223.
[18] Dávila-Jiménez, M.M., Elizalde-González, M.P. & Hernández-Montoya V. (2009). Performance of mango seed adsorbents in the adsorption of anthraquinone and azo acid dyes in single and binary aqueous solutions. *Bioresource Technology,* Vol. 100, No. 24, (December 2009), pp. (6199–6206), ISSN 0960-8524.
[19] Devi, R., Singh, V. & Kumar, A. (2008). COD and BOD reduction from coffee processing wastewater using Avocado peel carbon. *Bioresource Technology*, Vol. 99, No. 1, (April 2008), pp. (1853–1860), ISSN 0960-8524.
[20] Elizalde-González, M.P. & Hernández-Montoya, V. (2007). Characterization of mango pit as a raw material in the preparation of activated carbon for wastewater treatment. *Biochemical Engineering Journal,* Vol. 36, No. 3, (October 2007), pp. (230–238), ISSN 1369-703X.
[21] Elizalde-González, M.P. & Hernández-Montoya, V. (2008). Fruit seeds as adsorbents and precursors of carbon for the removal of anthraquinone dyes. *International Journal of Chemical Engineering*, Vol. 1, No. 2-3, pp. (243-253), ISSN 0974-5793.
[22] Elizalde-González, M.P. & Hernández-Montoya, V. (2009). Guava seed as adsorbent and as precursor of carbon for the adsorption of acid dyes. *Bioresource Technology*, Vol. 100, No. 7, (April 2009), pp. (2111–2117), ISSN 0960-8524.
[23] Elizalde-González, M.P. & Hernández-Montoya, V. (2009). Removal of acid orange 7 by guava seed carbon: A four parameter optimization study. *Journal of Hazardous Materials,* Vol. 168, No. 1, (August 2009), pp. (515 – 522), ISSN 0304-3894.
[24] Elizalde-González, M.P. & Hernández-Montoya, V. (2009). Use of wide-pore carbons to examine intermolecular interactions during adsorption of anthraquinone dyes from aqueous solution. *Adsorption Science & Technology,* Vol. 27, No. 5, (June 2009), pp. (447–459), ISSN 0263-6174.
[25] Elizalde-González, M.P. (2006). Development of non-carbonised natural adsorbents for removal of textile dyes. *Trends in Chemical Engineering,* Vol. 10, pp. (55–66), ISSN 0972-4478.
[26] Elizalde-González, M.P., Mattusch, J., Peláez-Cid, A.A. & Wennrich, R. (2007). Characterization of adsorbent materials prepared from avocado kernel seeds: Natural, activated and carbonized forms. *Journal of Analytical and Applied Pyrolysis*, Vol. 78, No. 1, (January 2007), pp. (185–193), ISSN 0165-2370.
[27] Elsayed, M.A., Hall, P.J. & Heslop, M.J. (2007). Preparation and structure characterization of carbons prepared from resorcinol-formaldehyde resin by CO_2 activation. *Adsorption*, Vol. 13, No. 3-4, pp. (299–306).
[28] Evans, M.J.B., MacDonald, J.A.F. & Halliop, E. (1999). The production of chemically-activated carbon. *Carbon*, Vol. 37, No. 2, (February 1999), pp. (269–274), ISSN 0008-6223.
[29] Gergova, K., Petrov, N. & Minkova, V. (1993). A comparison of adsorption characteristics of various activated carbons. *Journal of Chemical Technology and Biotechnology,* Vol. 56, No. 1, (April 2007 on line), pp. (77–82), ISSN 1097-4660.
[30] Gergova, K., Petrov, N. & Eser, S. (1994). Adsorption properties and microstructure of activated carbons produced from agricultural by-products by steam pyrolysis. *Carbon*, Vol. 32, No. 4, (May 1994), pp. (693–702), ISSN 0008-6223.

[31] Giraldo, L. & Moreno-Piraján, J.C. (2007). Calorimetric determinations of activated carbons in aqueous solution. *Journal of Thermal Analysis and Calorimetry*, Vol. 89, No.2, pp. (589–594), ISSN 1388-6150.
[32] Giraldo, L. & Moreno-Piraján, J. C. (2008). Pb^{2+} adsorption from aqueous solutions on activated carbons obtained from lignocellulosic residues. *Brazilian Journal of Chemical Engineering*, Vol. 25, No.1, (Jan./Mar. 2008), ISSN 0104-6632.
[33] Girgis, B.S. & Ishak, M.F. (1999). Activated carbon from cotton stalks by impregnation with phosphoric acid. *Materials Letters*, Vol. 39, No. 2, (April 1999), pp. (107–114), ISSN 0167-577X.
[34] Girgis, B.S., Yunis, S.S. & Soliman, A.M. (2002). Characteristics of activated carbon from peanut hulls in relation to conditions of preparation. *Materials Letters*, Vol. 57, No. 1, (November 2002), pp. (164–172), ISSN 0167-577X.
[35] Graham, N., Chen, X.G. & Jayaseelan, S. (2001). The potential application of activated carbon from sewage sludge to organic dyes removal. *Water Science and Technology*, Vol. 43, No. 2, pp. (245–252), ISSN 0273-1223.
[36] Guo, J., Gui, B., Xiang, S., Bao, X., Zhang, H., Lua, A.C. (2008). Preparation of activated carbons by utilizing solid wastes from palm oil processing mills. *Journal of Porous Mater*, Vol. 15, No. 5, (December 2003), pp. (535–540), ISSN 0165-2370.
[37] Gupta, V.K., Srivastava, S.K. & Mohan, D. (1997). Equilibrium uptake, sorption dynamics, process optimization, and column operation for the removal and recovery of malachite green from wastewater using activated carbon and activated slag. *Industrial and Engineering Chemistry Research*, Vol. 36, No.6, (June 1997), pp. (2207–2218), ISSN 0888-5885.
[38] Gupta, V.K., Mohan, D., Sharma, S. & Sharma M. (2000). Removal of basic dyes (Rhodamine B and Methylene Blue) from aqueous solution using bagasse fly ash. *Separation Science & Technology*, Vol. 35, No. 13, pp. (2097 - 2113), ISSN 0149-6395.
[39] Hameed, B.H., Din, A.T.M. & Ahmad, A.L. (2007). Adsorption of methylene blue onto bamboo-based activated carbon: Kinetics and equilibrium studies. *Journal of Hazardous Materials*, Vol. 141, No.3, (March 2007), pp. (819–825), ISSN 0304-3894.
[40] Hayashi, J., Horikawa, T., Takeda, I., Muroyama, K. & Ani, F.N. (2002). Preparing activated carbon from various nutshells by chemical activation with K2CO3. *Carbon*, Vol. 40, No. 13, (November 2002), pp. (2381-2386), ISSN 0008-6223.
[41] Hazourli, S., Ziati, M. & Hazourli A. (2009). Characterization of activated carbon prepared from lignocellulosic natural residue:-Example of date stones-. *Physics Procedia*, Vol. 2, No.3, pp. (1039–1043), ISSN 1875-3892.
[42] Heschel, W. & Klose, E. (1995). On the suitability of agricultural by-products for the manufacture of granular activated carbon. *Fuel*, Vol. 74, No. 12, (December 1995), pp. (1786–1791), ISSN 0016-2361.
[43] Hu, Z., Srinivasan, M.P. & Ni, Y. (2001). Novel activation process for preparing highly microporous and meso porous activated carbons. *Carbon*, Vol. 39, No. 6 (May 2001), pp. (877–886), ISSN 0008-6223.
[44] Iniesta, E., Sánchez, F., García, A.N. & Marcilla, A. (2001).Yields and CO2 reactivity of chars from almond shells obtained by a two heating step carbonisation process. Effect of different chemical pre-treatments and ash content. *Journal of Analytical and Applied Pyrolysis*, Vol. 58–59 (April 2001), pp. (983–994), ISSN 0165-2370.

[45] Jibril, B., Houache, O., Al-Maamari, R. & Al-Rashidi B. (2008). Effects of H3PO4 and KOH in carbonization of lignocellulosic material. Journal of Analytical and Applied Pyrolysis, Vol. 83, No. 2, (November 2008), pp. (151–156), ISSN 0165-2370.

[46] Juang, R.-S., Wu F.-C. & Tseng, R.-L. (2002). Characterization and use of activated carbons prepared from bagasses for liquid-phase adsorption. *Colloids and Surfaces A*, Vol. 201, No. 1-3, (March 2002),pp. (191–199), ISSN 0927-7757.

[47] Kalderis, D., Bethanis, S., Paraskeva, P. & Diamadopoulos, E. (2008). Production of activated carbon from bagasse and rice husk by a single-stage chemical activation method at low retention times. *Bioresource Technology*, Vol. 99, No. 15, (October 2008), pp. (6809–6816), ISSN 0960-8524.

[48] Kannan, N. & Sundaram, M.M. (2001). Kinetics and mechanism of removal of methylene blue by adsorption on various carbons–a comparative study. *Dyes and Pigments*, Vol. 51, No. 1, (October 2001), pp. (25–40), ISSN 0143-7208.

[49] Khalili, N.R., Campbell, M., Sandi, G. & Golas, J. (2000). Production of micro-and mesoporous activated carbon from paper mill sludge. I. Effect of zinc chloride activation. *Carbon*, Vol. 38, No. 14, (November 2000), pp. (1905–1915), ISSN 0008-6223.

[50] Ko, Y.G., Choi, U.S., Kim, J.S. & Park, Y.S. (2002). Novel synthesis and characterization of activated carbon fiber and dye adsorption modeling. *Carbon*, Vol. 40, No. 14, (November 2000), pp. (2661–2672), ISSN 0008-6223.

[51] Konstantinou, M. & Pashalidis, I. (2010). Competitive sorption of Cu(II) and Eu(III) ions on olive-cake carbon in aqueous solutions—a potentiometric study. *Adsorption*, Vol. 16, No. 3, (June 2010), pp. (167–171), ISSN 10450-010-9218-1.

[52] Kumar, M., Gupta, R.C. & Sharma, T. (1992). Influence of carbonisation temperature on the gasification of Acacia wood chars by carbon dioxide. *Fuel Processing Technology*, Vol. 32, No. 1-2, (November 1992), pp. (69-76), ISSN 0378-3820.

[53] Lillo-Ródenas, M.A., Lozano-Castelló, D., Cazorla-Amorós, D. & Linares-Solano, A. (2001). Preparation of activated carbons from Spanish anthracite, II. Activation by NaOH. *Carbon*, Vol. 39, No. 5, (April 2001), pp. (751–759), ISSN 0008-6223.

[54] Lozano-Castelló, D., Lillo-Ródenas, M.A., Cazorla-Amorós, D. & Linares-Solano, A. (2001). Preparation of activated carbons from Spanish anthracite, I. Activation by KOH. *Carbon*, Vol. 39, No. 5, (April 2001), pp. (741–749), ISSN 0008-6223.

[55] Lua, A.C. & Guo, J. (2000). Activated carbon prepared from oil palm stone by one-step CO2 activation for gaseous pollutant removal. *Carbon*, Vol. 38, No. 7, (June 2000), pp. (1089-1097), ISSN 0008-6223.

[56] Lussier, M.G., Shull, J.C. & Miller, D.J. (1994). Activated carbon from cherry stones. *Carbon*, Vol. 32, No. 8, (November 1994), pp. (1493–1498), ISSN 0008-6223.

[57] MacDonald, J.A.F. & Quinn, D.F. (1996). Adsorbents for methane storage made by phosphoric acid activation of peach pits. *Carbon*, Vol. 34, No. 9, (September 1996), pp. (1103–1108), ISSN 0008-6223.

[58] Malik, P.K. (2003). Use of activated carbons prepared from sawdust and rice husk for adsorption of acid dyes: a case study of Acid Yellow 36. *Dyes and Pigments*, Vol. 56, No. 3, (March 2003), pp. (239–249) ISSN 0143-7208.

[59] Marsh H. (Editor). (2001). *Activated carbon compendium*, Elsevier Science Ltd, ISBN: 0-08-044030-4, UK.

[60] Mizuta, K., Matsumoto, T., Hatate, Y., Nishihara, K. & Nakanishi, T. (2004). Removal of nitrate-nitrogen from drinking water using bamboo powder charcoal. *Bioresource Technology*, Vol. 95, No. 3, (December 2004), pp. (255–257), ISSN 0960-8524.

[61] Molina-Sabio, M., Rodríguez-Reinoso, F., Caturla, F. & Sellés, M.J. (1995). Porosity in granular carbons activated with phosphoric acid. *Carbon*, Vol. 33, No. 8, (August 1998), pp. (1105–1113), ISSN 0008-6223.

[62] Molina-Sabio, M., Rodríguez-Reinoso, F., Caturla, F. & Sellés, M.J. (1996). Development of porosity in combined phosphoric acid-carbon dioxide activation. *Carbon*, Vol. 34, No. 4, (April 1996), pp. (457–462), ISSN 0008-6223.

[63] Moreno-Castilla, C., Carrasco-Marín, F., López-Ramón, M.V. & Álvarez-Merino, M.A. (2001). Chemical and physical activation of olive-mill waste water to produce activated carbons. *Carbon*, Vol. 39, No. 9, (August 2001), pp. (1415-1420), ISSN 0008-6223.

[64] Mourão, P.A.M., Laginhas, C., Custódio, F., Nabais, J.M.V., Carrott, P.J.M. & Ribeiro-Carrott M.M.L. (2011). Influence of oxidation process on the adsorption capacity of activated carbons from lignocellulosic precursors. *Fuel Processing Technology*, Vol. 92, No. 2, (February 2011), pp. (241–246), ISSN 0378-3820.

[65] Nabais, J.M.V., Laginhas, C.E.C., Carrott, P.J.M. & Ribeiro-Carrott M.M.L. (2011). Production of activated carbons from almond shell. *Fuel Processing Technology*, Vol. 92, No. 2, (February 2011), pp. (234–240), ISSN 0378-3820.

[66] Namasivayam, C. & Kavitha, D. (2002). Removal of Congo Red from water by adsorption onto activated carbon prepared from coir pith, an agricultural solid waste. *Dyes and Pigments*, Vol. 54, No. 1, (July 2002), pp. (47–58), ISSN 0143-7208 .

[67] Ngernyen, Y., Tangsathitkulchai, C. & Tangsathitkulchai, M. (2006). Porous properties of activated carbon produced from Eucalyptus and Wattle wood by carbon dioxide activation. *Korean Journal of Chemical Engineering*, Vol. 23, No. 6, pp. (1046–1054), ISSN 0256-1115.

[68] Okada, K., Shimizu, Y.I., Kameshima, Y. & Nakajima, A. (2005). Preparation and Properties of Carbon/Zeolite Composites with Corrugated Structure. *Journal of Porous Materials*, Vol. 12, No. 4, pp. (281–291), ISSN 1380-2224.

[69] Olivares-Marín, M., Fernández-González, C., Macías-García, A. & Gómez-Serrano, V. (2006). Preparation of activated carbon from cherry stones by chemical activation with $ZnCl_2$. *Applied Surface Science*, Vol. 252, No. 17, (June 2006), pp. (5967–5971), ISSN 0169-4332.

[70] Olivares-Marín, M., Fernández-González, C., Macías-García, A. & Gómez-Serrano, V. (2006). Preparation of activated carbons from cherry stones by activation with potassium hydroxide. *Applied Surface Science*, Vol. 252, No. 17, (June 2006), pp. (5980–5983), ISSN 0169-4332.

[71] Ould-Idriss, A., Stitou, M., Cuerda-Correa, E.M., Fernández-González, C.A., Macías-García, A., Alexandre-Franco, M.F. & Gómez-Serrano V. (2011). Preparation of activated carbons from olive-tree wood revisited. II. Physical activation with air. *Fuel Processing Technology*, Vol. 92, No. (July 2010), pp. (266–270), ISSN 0378-3820.

[72] Phan, N.H., Rio, S., Faur, C., Le Coq L., Le Cloirec, P. & Nguyen, T.H. (2006). Production of fibrous activated carbons from natural cellulose (jute, coconut) fibers for water treatment applications. *Carbon*, Vol. 44, No. 12, (October 2006), pp. (2569–2577), ISSN 0008-6223.

[73] Puziy, A.M., Poddubnaya, O.I., Martínez-Alonso, A., Suárez-García, F. & Tascón, J.M.D. (2005). Surface chemistry of phosphorus-containing carbons of lignocellulosic origin. *Carbon*, Vol. 43, No. 14, (November 2005), pp. (2857-2868), ISSN 0008-6223.

[74] Qian, Q., Machida, M., Aikawa, M. & Tatsumoto, H. (2008). Effect of $ZnCl_2$ impregnation ratio on pore structure of activated carbons prepared from cattle manure compost: Application of N_2 adsorption desorption isotherms. *Journal of Material Cycles Waste Management*, Vol. 10, No. 1, pp. (53–61), ISSN 1438-4957.

[75] Rajeshwarisivaraj, Sivakumar, S., Senthilkumar, P. & Subburam, V. (2001). Carbon from Cassava peel, an agricultural waste, as an adsorbent in the removal of dyes and metal ions from aqueous solution. *Bioresource Technology*, Vol. 80, No. 3, (December 2001), pp. (233–235), ISSN 0960-8524.

[76] Robau-Sánchez, A., Aguilar-Elguézabal, A. & De La Torre-Saenz, L. (2001). CO_2 activation of char from quercus agrifolia wood waste. *Carbon*, Vol. 39, No. 9, (August 2001), pp. (1367–1377) ISSN 0008-6223.

[77] Robinson, T., McMullan, G., Marchant, R & Nigam, P. (2001). Remediation of dyes in textile effluent: a critical review on current treatment technologies with a proposed alternative, *Bioresource Technology*, Vol. 77, No. 3, (May 2001), pp. (247–255), ISSN 0960-8524.

[78] Rodríguez-Mirasol, J., Cordero, T. & Rodríguez J.J. (1993). Preparation and characterization of activated carbons from eucalyptus kraft lignin. *Carbon*, Vol. 31, No. 1, (January 1993), pp. (87–95), ISSN 0008-6223.

[79] Rodríguez-Reinoso, R. & Molina-Sabio, M. (1992), Activated carbons from lignocellulosic materials by chemical and/or physical activation: An overview. *Carbon*, Vol. 30, No. 7, (October 1992), pp. (1111–1118), ISSN 0008-6223.

[80] Salame, I.I. & Bandosz, T.J. (2001). Surface chemistry of Activated Carbons: Combining the results of Temperature-Programmed Desorption, Boehm and Potentiometric Titrations. *Journal of Colloids and Interface Science*, Vol. 240, No. 1, (August 2001), pp. (252–258), ISSN 0021-9797.

[81] Salunkhe, D.K., & Kadam, S.S. (Editors). (1995). *Handbook of fruit science and technology, production, composition, storage and processing*, Marcel Dekker, Inc., ISBN: 0-8247-9643-8, USA.

[82] Senthilkumaar, S., Kalaamani, P., Porkodi, K., Varadarajan, P.R. & Subburaam, C.V. (2006). Adsorption of dissolved Reactive red dye from aqueous phase onto activated carbon prepared from agricultural waste. *Bioresource Technology*, Vol. 97, No. 14, (September 2006), pp. (1618–1625), ISSN 0960-8524.

[83] Sun, R.Q., Sun, L.B., Chun, Y. & Xu, Q.H. (2008). Catalytic performance of porous carbons obtained by chemical activation. *Carbon*, Vol. 46, No. 13, (November 2008), pp. (1757–1764) ISSN 0008-6223.

[84] Swarnalatha, S., Ganesh-Kumar, A. & Sekaran, G. (2009). Electron rich porous carbon/silica matrix from rice husk and its characterization. *Journal of Porous Mater*, Vol. 16, No. 3, pp. (239–245), ISSN 1380-2224.

[85] Tay, J.H., Chen, X.G., Jeyaseelan, S. & Graham, N. (2001). Optimising the preparation of activated carbon from digested sewage sludge and coconut husk. *Chemosphere*, Vol. 44, No. 1, (July 2001), pp. (45–51), ISSN 0045-6535.

[86] Teng, H., Ho, J.A. & Hsu, Y.F. (1997). Preparation of activated carbons from bituminous coals with CO_2 activation-influence of coal oxidation. *Carbon*, Vol. 35, No. 2, (February 1997), pp. (275–283), ISSN 0008-6223.

[87] Teng, H., Yeh, T.S. & Hsu, L.Y. (1998). Preparation of activated carbon from bituminous coals with phosphoric acid activation. *Carbon*, Vol. 36, No. 9, (September 1998), pp. (1387–1395), ISSN 0008-6223.

[88] Teng, Y.C., Lin, L.Y. & Hsu, H. (2000). Production of activated carbons from pyrolysis of waste tires impregnated with potassium hydroxide. *Journal of Air Waste Management Association*, Vol. 50, (November 2000), pp. (1940–1946), ISSN 1047-3289.

[89] Toles, C.A., Marshall, W.E., Johns, M.M. (1997). Granular activated carbons from nutshells for the uptake of metals and organic compounds. *Carbon*, Vol. 35, No. 9, (September 1997), pp. (1407-1414), ISSN 0008-6223.

[90] Tsai, W.T., Chang, C.Y. & Lee, S.L. (1997). Preparation and characterization of activated carbons from corn cob. *Carbon*. Vol. 35, No. 8, (November 1997), pp. (1198–1200), ISSN 0008-6223.

[91] Tsai, W.T., Chang, C.Y. & Lee, S.L. (1998). A low cost adsorbent from agricultural waste corn cob by zinc chloride activation. *Bioresource Technology*, Vol. 64, No. 3, (June 1998), pp. (211–217), ISSN 0960-8524.

[92] Tsai, W.T., Chang, C.Y., Lin, M.C., Chien, S.F., Sun, H.F. & Hsieh, M.F. (2001a). Adsorption of acid dye onto activated carbons prepared from agricultural waste bagasse by $ZnCl_2$ activation. *Chemosphere*, Vol. 45, No. 1, (October 2001), pp. (51–58), ISSN 0045-6535.

[93] Tsai, W.T., Chang, C.Y., Wang, S.Y., Chang, C.F., Chien, S.F. & Sun, H.F. (2001b). Preparation of activated carbons from corn cob, catalyzed by potassium salts and subsequent gasification with CO_2, *Bioresource Technology*, Vol. 78, No. 2, (June 2001), pp. (203–208), ISSN 0960-8524.

[94] Tsai, W.T., Chang, C.Y., Wang, S.Y., Chang, C.F., Chien, S.F. & Sun, H.F. (2001c). Cleaner production of carbon adsorbents by utilizing agricultural waste corn cob. *Resources, Conservation and Recycling*, Vol. 32, No. 1, (May 2001), pp. (43–53), ISSN 0921-3449.

[95] Tseng, R.-L. & Tseng, S.-K. (2005). Pore structure and adsorption performance of the KOH-activated carbons prepared from corncob. *Journal of Colloid and Interface Science*, Vol. 287, No. 2, (July 2005), pp. (428–437), ISSN 0021-9797.

[96] Valente-Nabais, J.M., Gomes, J.A., Suhas, Carrott, P.J.M., Laginhas, C. & Roman, S. (2009). Phenol removal onto novel activated carbons made from lignocellulosic precursors: influence of surface properties, *Journal of Hazardous Materials*, Vol. 167, No. 1-3, (August 2009), pp. (904–910), ISSN 0304-3894.

[97] Vargas, A.M.M., Cazetta, A.L., Garcia, C.A., Moraes, J.C.G., Nogami, E.M., Lenzi, E., Costa, W.F. & Almeida, V.C. (2011). Preparation and characterization of activated carbon from a new raw lignocellulosic material: flamboyant (*Delonix Regia*) pods. *Journal of Environmental Management*, Vol. 92, No. 1, (January 2011), pp. (178-184), ISSN 0301-4797.

[98] Vargas, J.E., Giraldo, L. & Moreno-Piraján, J.C. (2010). Preparation of activated carbons from seeds of Macuna mutisiana by physical activation with steam. *Journal of Analytical and Applied Pyrolisis*, Vol. 89, No. 2, (November 2010), pp. (307-312), ISSN 0165-2370.

[99] Warhurst, A.M., Fowler, G.D., McConnachie, G.L. & Pollard, S.J.T. (1997). Pore structure and adsorption characteristics of steam pyrolysis carbons from *Moringa oleifera*. *Carbon*, Vol. 35, No. 8, (August 1997), pp. (1039–1045), ISSN 0008-6223.

[100] Wu, F.C., Tseng, R.L. & Juang, R.S. (1999). Pore structure and adsorption performance of the activated carbons prepared from plum kernels. *Journal of Hazardous Materials*, Vol. 69, No. 3, (November 1999), pp. (287–302), ISSN 0304-3894.

[101] Wu, F.C., Tseng, R.L. & Juang, R.S. (2001). Adsorption of dyes and phenol from water on the activated carbons prepared from corncob wastes. *Environmental Technology*, Vol. 22, No. 2, (February 2001), pp. (205–213), ISSN 0959-3330.

[102] Yavuz, R.; Akyildiz, H.; Karatepe, N. & Çetinkaya, E. (2010). Influence of preparation conditions on porous structures of olive stone activated by H3PO4. Fuel Processing Technology, Vol. 91, No. 1, (January 2010), pp. (80–87), ISSN 0378-3820.
[103] Zhang, H., Yan, Y. & Yang, L. (2010). Preparation of activated carbon from sawdust by zinc chloride activation. *Adsorption*, Vol. 16, No. 3, (August 2010), pp. (161–166).
[104] Zuo, S., Yang, J. & Liu, J. (2010). Effects of the heating history of impregnated lignocellulosic material on pore development during phosphoric acid activation. *Carbon*, Vol. 48, No. 11, (September 2010), pp. (3293–3295), ISSN 0008-6223.

2

Thermal Treatments and Activation Procedures Used in the Preparation of Activated Carbons

Virginia Hernández-Montoya, Josafat García-Servin and José Iván Bueno-López
Instituto Tecnológico de Aguascalientes
México

1. Introduction

The preparation of activated carbons (ACs) generally comprises two steps, the first is the carbonization of a raw material or precursor and the second is the carbon activation. The carbonization consists of a thermal decomposition of raw materials, eliminating non-carbon species and producing a fixed carbon mass with a rudimentary pore structure (very small and closed pores are created during this step). On the other hand, the purpose of activation is to enlarge the diameters of the small pores and to create new pores and it can be carried out by chemical or physical means. During chemical activation, carbonization and activation are accomplished in a single step by carrying out thermal decomposition of the raw material impregnated with certain chemical agents such as H_3PO_4, H_2SO_4, HNO_3, NaOH, KOH and $ZnCl_2$ (Hu et al., 2001; Mohamed et al., 2010). Physical or thermal activation uses an oxidizing gas (CO_2, steam, air, etc.) for the activation of carbons after carbonization, in the temperature range from 800 to 1100 °C. The carbonization can be carried out using tubular furnaces, reactors, muffle furnace and, more recently, in glass reactor placed in a modified microwave oven (Foo & Hameed, 2011; Tongpoothorn et al., 2011; Vargas et al., 2010).

Nowadays, the raw materials more used in the preparation of carbons are of lignocellulosic origin. Wood and coconut shells are the major precursors and responsible for the world production of more than 300, 000 tons/year of ACs (Mourão et al., 2011). However, the precursor selection depends of their availability, cost and purity, but the manufacturing process and the application of the product are also important considerations (Yavuz et al., 2010). Figure 1 shows the number of publications studied in this chapter, related with the preparation of activated carbons from lignocellulosic materials in last two decades. A clear trend can be observed: the number of works increased in the years from 2000 to 2010. The obtained carbons were mainly employed in the removal of water pollutants.

In the present chapter the principal methods used in the preparation of activated carbons from lignocellulosic materials by chemical and physical procedures are discussed. An analysis of the experimental conditions used in the synthesis of ACs has been made attending to the carbon specific surface area. Also the advantages and disadvantages of each method are discussed.

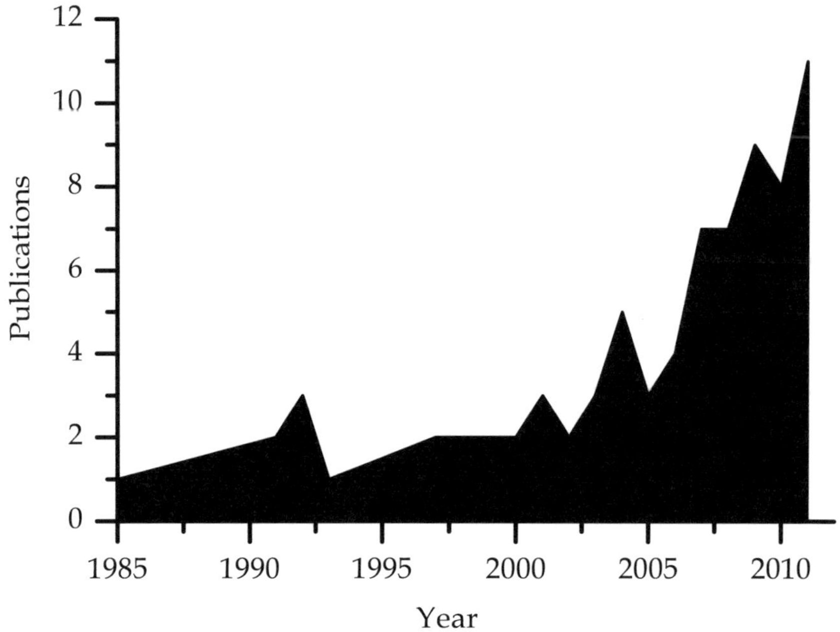

Figure 1. Number of publications related with the preparation of activated carbons from lignocellulosic precursors in the last two decades

2. Preparation of activated carbons

The preparation of ACs from lignocellulosic materials involved two processes, the carbonization and the activation, which can be performed in one or two steps depending on the activation method (physical or chemical, respectively). Specifically, when the carbonization is carried out in an inert atmosphere the process is called *pyrolysis*. According to the literature, the pyrolysis of lignocellulosic materials as coconut shells, olive stones, walnut shells, etc., gives rise to three phases: the char, oils (tars) and gases. The relative amount of each phase is a function of parameters such as temperature of pyrolysis, nitrogen flow rate and heating rate. For example, slow heating rates promote high yields of the carbon residue while flash pyrolysis is recommended for high liquid (oil) ratios (Mohamed et al., 2010).

During the pyrolysis of lignocellulosic precursors, a rudimentary porosity is obtained on the char fraction as a consequence of the release of most of the non-carbon elements such as hydrogen, oxygen and nitrogen in form of gases and tars, leaving a rigid carbon skeleton formed by aromatic structures.

There are two conventional methods for activating carbons: physical (or thermal) and chemical activation. During the chemical activation, the precursor is first impregnated or physically mixed with a chemical compound, generally a dehydrating agent. The impregnated carbon or the mixture is then heated in an inert atmosphere (Moreno-Castilla et al., 2001). On the other hand, during a physical activation process the lignocellulosic precursor is carbonized under an inert atmosphere, and the resulting carbon is subjected to a partial and controlled gasification at high temperature (Rodriguez-Reinoso & Molina-Sabio, 1992).

In the following sections the principal characteristics of the procedures used in the preparation of activated carbons from lignocellulosic precursors by physical and chemical methods are described.

2.1 Chemical activation

The carbonization step and the activation step simultaneously progress in the chemical activation (Hayashi et al., 2002a). In this case, the lignocellulosic precursor is treated primarily with a chemical agent, such as H_3PO_4, H_2SO_4, HNO_3, NaOH, KOH or $ZnCl_2$ by impregnation or physical mixture and the resulting precursor is carbonized at temperatures between 400 and 800 °C under a controlled atmosphere. The function of the dehydrating agents is to inhibit the formation of tar and other undesired products during the carbonization process. Also, the pore size distribution and surface area are determined by the ratio between the mass of the chemical agent and the raw material. Besides, activation time, carbonization temperature and heating rate are important preparation variables for obtaining ACs with specific characteristics (Mohamed et al., 2010). The effects of all these parameters in the textural characteristics of ACs employing different activating agents are discussed in the following sections.

2.1.1 Phosphoric acid (H_3PO_4)

In the last 20 years, the activation of lignocellulosic materials with H_3PO_4 has become an increasingly used method for the large-scale manufacture of ACs because the use of this reagent has some environmental advantages such as ease of recovery, low energy cost and high carbon yield. H_3PO_4 plays two roles during the preparation of ACs: i) H_3PO_4 acts as an acid catalyst to promote bond cleavage, hydrolysis, dehydration and condensation, accompanied by cross-linking reactions between phosphoric acid and biopolymers; ii) H_3PO_4 may function as a template because the volume occupied by phosphoric acid in the interior of the activated precursor is coincident with the micropore volume of the activated carbon obtained (Zuo et al., 2009).

The chemical and physical properties of ACs obtained by chemical activation with H_3PO_4 are affected by the experimental conditions of preparation such as acid concentration, time of activation, impregnation ratio, carbonization temperature and heating rate. Also some recent works have shown that the atmosphere used in the carbonization process has an obvious effect on the physicochemical properties of ACs (Zuo et al., 2009). Table 1 collects some experimental conditions used in the preparation of activated carbons from lignocellulosic materials using N_2 as activation atmosphere.

Precursor	H_3PO_4 (%)	Impregnation ratio	Activation temperature (°C)	Heating rate (°C min^{-1})	Reference
Avocado kernel seed	85	6	800	5	Elizalde-González et al. (2007)
China fir	50	4.6	475	5	Zuo et al. (2009)
Coconut Fibers	30	4	900	20	Phan et al. (2006)
Fruit stones	60	1.02	800	-	Puziy et al. (2005)
Jackfruit peel waste	85	4	550	-	Prahas et al. (2008)
Jute	30	4	900	20	Phan et al. (2006)
Licorice residues	89	1.5	400	2.5	Kaghazchi et al. (2010)
Oil palm shell	85	0.09	450	5	Arami-Niya et al. (2011)
Olive Stone	50	2	400	5	Yavuz et al. (2010)
Olive waste	75	2.4	500	10	Moreno-Castilla et al. (2001)
Pecan shell	50	-	450	-	Ahmedna et al. (2000)
Pine Wood	85	1.5	400	-	Hared et al. (2007)
Pistachio-nut shells	89	0.5	400	5	Kaghazchi et al. (2010)
Sea-buckthorn stones	85	0.5	550	10	Mohammadi et al. (2010)
Stem of date palm	85	5	600	10	Jibril et al. (2008)
Tea plant	85	3	350	-	Yagmur et al. (2008)

Table 1. Experimental conditions of activated carbons obtained by chemical activation with H_3PO_4 using different lignocellulosic precursors

In most of the cited papers, the concentration of acid is greater than 50% (w/w) and the activation temperature for 75 % of these studies is between 350 and 600 °C (see Table 1). Figure 2 shows the specific surface area calculated by the Brunauer, Emmett and Teller method (S_{BET}) of the ACs prepared in the contributions collected in Table 1. Carbons obtained with the highest phosphoric impregnation ratio (China Fir and avocado kernel seeds) are the materials with the largest S_{BET} (1785 and 1802 m^2 g^{-1}). Additionally, the carbon obtained from Oil palm shell and activated using a rather low impregnation ratio (0.09) was one of the materials with a lower specific surface area (356 m^2 g^{-1}).

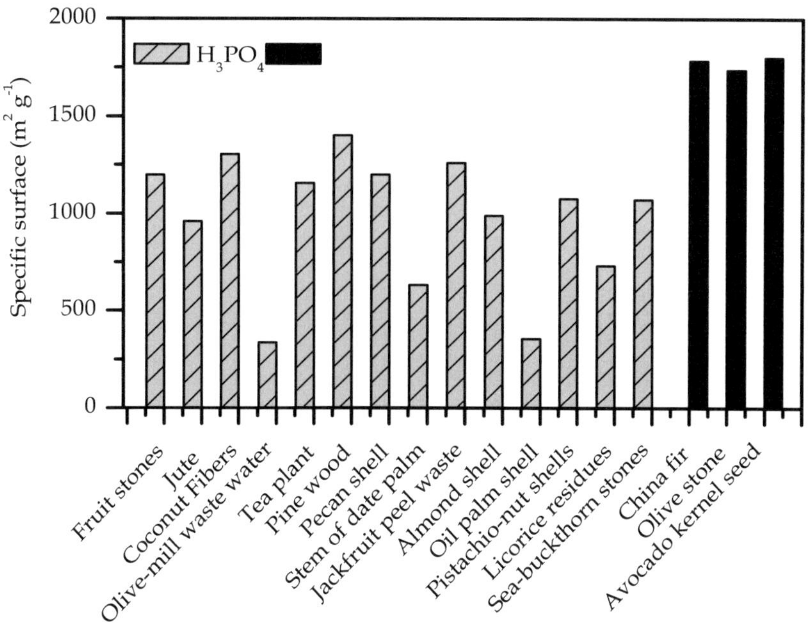

Figure 2. Specific surface area of activated carbons obtained by chemical activation of lignocellulosic materials with H_3PO_4 (black bars: ACs with greater S_{BET})

2.1.2 Zinc Chloride ($ZnCl_2$)

Chemical activation of lignocellulosic materials with $ZnCl_2$ leads to the production of activated carbons with good yield a well-developed porosity in only one step. Impregnation with $ZnCl_2$ first results in degradation of the material and, on carbonization, produces dehydration that results in charring and aromatization of the carbon skeleton and creation of the pore structure (Caturla et al., 1991). In this case, the precursor is impregnated with a concentrated $ZnCl_2$ solution during a given contact time, followed by evaporation of the solution and, finally, the precursor is carbonized in an inert atmosphere and thoroughly washed to extract the excess of $ZnCl_2$. The amount of $ZnCl_2$ incorporated in the precursor and the temperature of heat treatment are the two variables with a direct incidence in the development of the porosity. Table 2 shows the experimental conditions used in the preparation of ACs by chemical activation with $ZnCl_2$ using N_2 as activation atmosphere.

The specific surface areas of the carbons reported in the papers of Table 2 are shown in Figure 3. Carbons obtained using the highest impregnation ratios (2 and 2.5) and an activation temperature of 800 °C are the materials with the largest S_{BET} (Caturla et al., 1991; Hu et al., 2001). The carbon obtained from coconout shells reaches an S_{BET} value of 2400 m^2 g^{-1}, whereas for the carbon prepared from peach stones the S_{BET} was 2000 m^2 g^{-1}. Other carbons prepared from coconut shells using an impregnation ratio of 1 and an activation temperature of 500 °C show lower specific surface areas (1200 m^2 g^{-1}). In any case, all the carbons prepared by chemical activation with $ZnCl_2$ attain S_{BET} greater than 750 m^2 g^{-1} (Azevedo et al., 2007). The principal disadvantage of this activation is the environmental risks related to zinc compounds.

Precursor	Impregnation ratio (IR)	Activation temperature (°C)	Heating rate (°C min^{-1})	Reference
Almond shells	2	600	-	Torregrosa &Martín (1991)
Cherry stones	3	500	10	Olivares-Marín et al. (2006)
Coconut shells	1	500	4	Azevedo et al. (2007)
Coconut shells*	2	800	10	Hu et al. (2001)
Coffee residue	1	600	10	Boudrahem et al. (2009)
Coir Pith	1	700	-	Namasivayam & Kadirvelu (1997)
Hypnea valentiae	-	800	10	Aravindhan et al. (2009)
Licorice residues	1	500	2.5	Kaghazchi et al. (2010)
Oil palm shell	-	500	5	Arami-Niya et al. (2011)
Oil palm shells	-	500	10	Arami-Niya et al. (2010)
Peach stones	2.5	800	-	Caturla et al. (1991)
Pistachio-nut shells	1.5	500	5	Kaghazchi et al. (2010)
Sargassum longifolium	-	800	10	Aravindhan et al. (2009)
Sea-buckthorn stones	0.5	550	-	Mohammadi et al. (2010)
Walnut shells	2	450	5	Yang & Qiu (2010)

Table 2. Experimental conditions of activated carbons obtained by chemical activation with $ZnCl_2$ using different lignocellulosic precursors

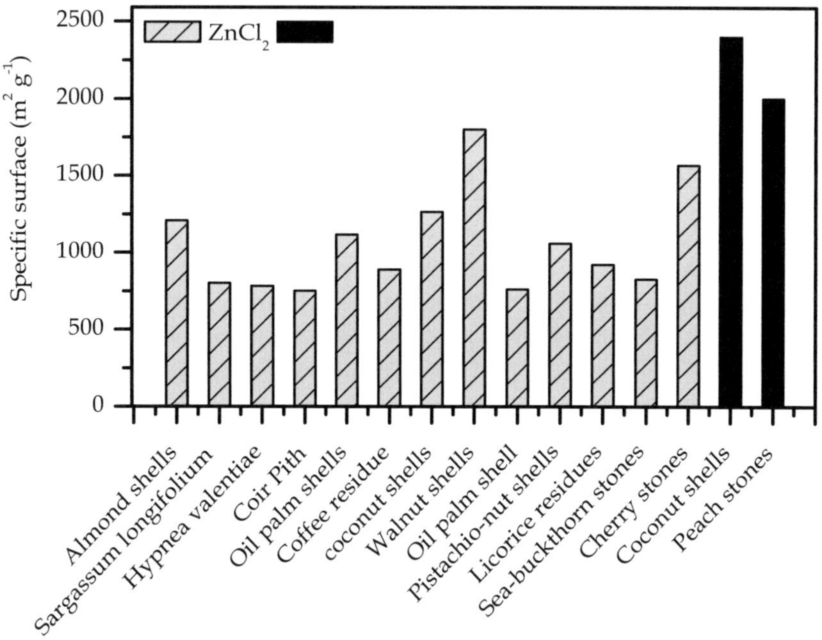

Figure 3. Specific surface area of activated carbons obtained by chemical activation of lignocellulosic materials with $ZnCl_2$ (black bars: ACs with greater S_{BET})

2.1.3 Alkalis

Alkaline hidroxides (KOH, NaOH) and carbonates (K_2CO_3, Na_2CO_3) have been used as activation reagents in the preparation of activated carbons with high specific surface. In general terms, chemical activation by KOH and NaOH consists in a solid-solid or solid-liquid reaction involving the hydroxide reduction and carbon oxidation to generate porosity (Adinata et al., 2007). The activation with KOH was first reported in the late 1970s by AMOCO Corporation; since then many studies have been devoted to the preparation of ACs by chemical activation with KOH (Lua & Yang, 2004). In this context, two procedures have been used. The carbon precursor can be mixed with powder of KOH or impregnated with a concentrated solution of KOH and then the solid mixture or impregnated precursor is thermally treated under nitrogen (Bagheri & Abedi, 2009; Moreno-Castilla et al., 2001). Alternatively, the preparation of ACs by alkaline activation is made in two steps, in which the precursor is first pyrolyzed and the obtained carbon is activated with a solution of KOH (Bagheri & Abedi, 2009) or with pellets of KOH and finally thermally treated again. The activation step can be conducted in a glass reactor placed in a modified micro wave oven with a frequency of 2.45 GHz (Foo & Hameed, 2011).

Sodium hidroxide has been also shown to be more interesting activation agent due to the possibility of reducing chemical activation costs and environmental load when compared with KOH activation (Tongpoothorn et al., 2011). The activation procedure with NaOH is similar to KOH (Tseng, 2007; Vargas et al., 2011).

In general, the preparation of ACs by chemical activation with KOH and NaOH allows to obtain carbons with high specific surface areas (>1000 m² g⁻¹). However, KOH and NaOH are corrosive and deleterious chemicals (Hayashi et al., 2002a). For this reason, recent studies have proposed the preparation of activated carbons by chemical activation with K_2CO_3 in one step, in which the lignocellulosic materials is impregnated with a K_2CO_3 solution and finally the impregnated precursor is thermally treated. K_2CO_3 is a not deleterious reagent and it is broadly used for food additives (Hayashi et al., 2002a).

Table 3 summarizes the experimental conditions used in the preparation of ACs from lignocellulosic materials by chemical activation with NaOH, KOH and K_2CO_3. Carbons obtained by activation with NaOH are the materials showing higher S_{BET} (see Figure 4), for example, the carbon obtained from flamboyant exhibiting a S_{BET} near to 2500 m^2 g^{-1}. Also, the activation with K_2CO_3 renders carbons with a competitive S_{BET} (between 1200 and 1800 m^2 g^{-1}) compared with those obtained by activation with KOH or NaOH.

Other interesting observation is that the specific surface areas of two ACs obtained from pistachio nut shells activated with KOH and treated in two different thermal configurations (a conventional electric oven and a modified microwave oven), were very similar (700 and 796 m^2 g^{-1}), thus suggesting that the two methods (conventional and non-conventional) are effective for the preparation of ACs.

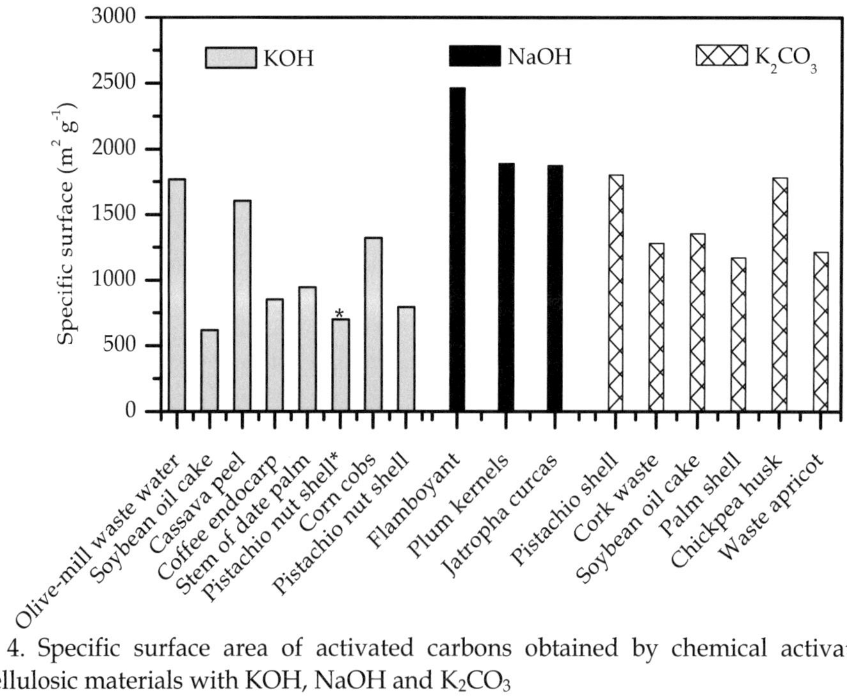

Figure 4. Specific surface area of activated carbons obtained by chemical activation of lignocellulosic materials with KOH, NaOH and K_2CO_3

Precursor	Activating state	IR	Carbonization temperature (°C)	Heating rate (°C min^{-1})	Reference
			KOH		
Cassava peel	-	2.5	750	10	Sudaryanto et al. (2006)
Coffee endocarp	Powder	1:2	850	5	Nabais et al. (2008)
Corn cobs	Solution	2	550	10	Bagheri & Abedi (2009)
Olive-mill waste water	Solution	2	800	10	Moreno-Castilla et al. (2001)
Pistachio nut shell	Pellets	0.5	300	10	Lua & Yang (2004)
Pistachio nut shell*	Solution	1.75	Frequency of 2.45 GHz.	600 W	Foo & Hameed (2011)
Soybean oil cake	Solution	0.95	800	5	Tay et al. (2009)
Stem of date palm	Solution	3	600	50	Jibril et al. (20089
NaOH					
Flamboyant	Pellets	3	700	20	Vargas et al. (2011)
Jatropha curcas	Solution	4	400	-	Tongpoothorn et al. (2011)
Plum kernels	Solution	4	780	-	Tseng (2007)
K_2CO_3					
Chickpea husk	Solution	-	800	10	Hayashi et al. (2002b)
Cork waste	Solution	3	800	10	Carvalho et al. (2004)
Palm shell	Solution	2	800	10	Adinata et al. (2007)
Pistachio shell	Solution	-	800	10	Hayashi et al. (2002a)
Soybean oil cake	Solution	1	800	5	Tay et al. (2009)
Waste apricot	Solution	1	900	10	Erdoğan et al. (2005)

Table 3. Experimental conditions of activated carbons obtained by chemical activation with NaOH and KOH using different lignocellulosic precursors

2.2 Physical or thermal activation

In a physical activation process, the lignocellulosic precursor is carbonized under an inert atmosphere, and the resulting carbon is subjected to a partial and controlled gasification at high temperature with steam, carbon dioxide, air or a mixture of these (Rodriguez-Reinoso & Molina-Sabio, 1992). Steam and CO_2 are the two activating gases more used in the physical activation of carbons. According to the literature, steam or CO_2 react with the carbon structures to produce CO, CO_2, H_2 or CH_4. The degree of activation is normally referred to as "burn-off" and it is defined as the weight difference between the carbon and the activated carbon divided by the weight of the original carbon on dry basis according with the following equation,

$$Burn\ off = \frac{W_0 - W_1}{W_0} X100\% \tag{1}$$

where W_0 is the weight of the original carbon and W_1 refers to the mass of the activated carbon. The use of CO_2 during the activation process of a carbon material develops narrow micropores, while steam widens the initial micropores of the carbon. At high degrees of burn-off, steam generates activated carbons with larger meso and macropore volumes than those prepared by CO_2. Consequently, CO_2 creates activated carbons with larger micropore volumes and narrower micropore size distributions than those activated by steam (Mohamed et al., 2010)

Tables 4 and 5 show the experimental conditions used in the preparation of activated carbons from lignocellulosic materials by physical activation with CO_2, steam and steam-N_2 admixtures. Normally, in these experiments the lignocellulosic precursor is carbonized in an inert atmosphere (N_2) at temperatures ranging from 400 to 950 °C to produce carbons with rudimentary pore structures. These carbons are then activated with the selected gasification agent at temperatures around 800-1000 °C to produce the final activated carbons.

Some additional studies combine the thermal or physical activation with chemical activation (also known as physicochemical activation, Table 6). Normally, physicochemical activation is performed by changing the activation atmosphere of the chemical activation by a gasification atmosphere (i.e., steam) at higher temperatures. In other cases, the chemical activation is carried out directly under the presence of a gasifying agent. The combination of both types of carbon activation renders ACs with textural and chemical properties which are different from those obtained by any of the activations alone. For example, steam reduces the occurrence of heteroatoms into the carbon structures. Also, combination of oxidizing reagents in the liquid phase (i.e., nitric or sulfuric acids) with gasification agents improves the development of porosity on the final carbons.

Figure 5 shows the specific surface area of activated carbons obtained by physical and physiochemical activation according with the experimental conditions cited in Tables 4, 5 and 6. In general, the ACs obtained by physical activation with CO_2 show a higher specific surface area that those obtained by activation with steam. Additionally, the ACs obtained by physical activation with CO_2 using high heating rates (20 °C min^{-1}) are the adsorbents showing lower S_{BET} (Corncob, Bagasse bottom ash and Sawdust fly ash).

Precursor	Carbonization Conditions			Activation conditions				Reference
	Temperature (°C)	Heating rate (°C min^{-1})	Atm.	Temperature (°C)	Heating rate (°C min^{-1})	Atm.	Flow (cm^3 min^{-1})	
Almond shell	400	10	N_2	800	10	CO_2	85	Mourão et al. (2011)
Almond shell	400	10	N_2	800	10	CO_2	85	Nabais et al. (2011)
Bagasse bottom ash	500	20	N_2	800	20	CO_2	200	Aworn et al. (2008)
Coconut Fibers	950	-	N_2	950	-	CO_2	250	Phan et al. (20069
Coffee endocarp	700	5	N_2	800	5	CO_2	85	Nabais et al. (2008)
Corncob	500	20	N_2	800	20	CO_2	200	Aworn et al. (2008)
Eucalyptus kraft lignin	350	10	N_2	800	10	CO_2	150	Rodriguez-Mirasol et al. (19939
Jute	950	-	N_2	950	-	CO_2	250	Phan et al. (2006)
Macadamia nut-shell	500	20	N_2	800	20	CO_2	200	Aworn et al. (2008)
Oil-palm-shell	600	10	N_2	900	10	CO_2	-	Lua et al. (2006)
Olive-mill waste	350	-	Air	850	-	CO_2	300	Moreno-Castilla et al. (2001)
Pistachio-nut shells	500	10	N_2	900	10	CO_2	100	Lua et al. (2004)
Pistachio-nut shells	500	10	N_2	800	10	CO_2	100	Yang & Lua (2003)
Sawdust fly ash	500	20	N_2	800	20	CO_2	200	Aworn et al. (2008)
Vine shoots	400	10	N_2	800	10	CO_2	85	Mourão et al. (2011)

Table 4. Experimental conditions of activated carbons obtained from different lignocellulosic precursors by physical activation with CO_2

Precursor	Carbonization Conditions			Activation conditions				Reference
	Temperature (°C)	Heating rate (°C min^{-1})	Atm.	Temperature (°C)	Heating rate (°C min^{-1})	Atm.	Flow (cm^3 min^{-1})	
Almond shell	600	-	N_2	850	-	Steam -N_2	150	González et al. (2009)
Almond tree pruning	600	-	N_2	850	-	Steam -N_2	150	González et al. (2009)
Date stones	700	10	N_2	700	10	Steam	100	Bouchelta et al. (2008)
M. oleiferu seed	-	-	-	800	20	Steam	-	Warhurs et al. (1997)
Olive stone	600	-	N_2	850	-	Steam -N_2	150	González et al. (2009)
Walnut shell	600	-	N_2	850	-	Steam -N_2	150	González et al. (2009)

Table 5. Experimental conditions of activated carbons obtained from various lignocellulosic precursors by physical activation with steam

Precursor	Activating agent	Temperature (°C)	Heating rate (°C min^{-1})	Atm.	Flow (cm^3 min^{-1})	Reference
Date stones	H_3PO_4	600	-	Steam	-	Hazourli et al. (2009)
Date stones	HNO_3	600	-	Steam	-	Hazourli et al. (2009)
Olive stones	$CaCl_2$	800	-	CO_2	100	Juárez-Galán et al. (2009)
Sugarcane bagasse	H_2SO_4	160	10	Air	2000	Valix et al. (2004)

Table 6. Experimental conditions of activated carbons obtained from different lignocellulosic precursors by physiochemical activation

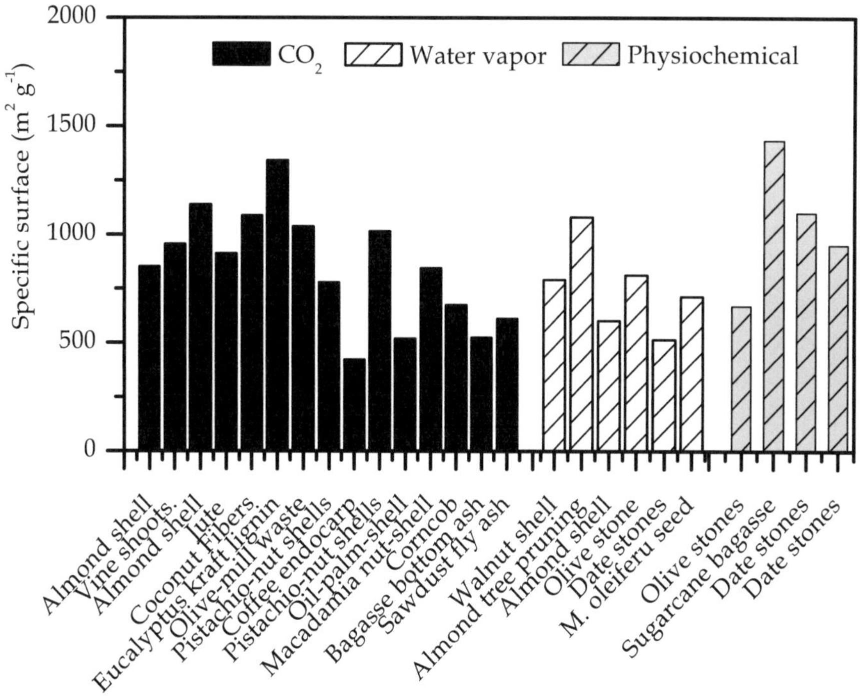

Figure 5. Specific surface of activated carbons obtained by physical activation with CO_2 and water vapor and by physiochemical activation

3. Analysis of the methods used in the preparation of ACs

The advantages and drawbacks of the different types of carbon activation are discussed in the following points.

3.1 Chemical method

Advantages

- ✓ Activated carbons are obtained in one step
- ✓ Shorter activation times
- ✓ Lower temperatures of pyrolysis (600 an 800 °C)
- ✓ Better control of textural properties
- ✓ High yield
- ✓ High surface area of the ACs
- ✓ Well-developed microporosity
- ✓ Narrow micropore size distributions
- ✓ Reduction of the mineral matter content

Disadvantages

- ✓ Corrosiveness of the process
- ✓ Requires a washing stage
- ✓ Inorganic impurities
- ✓ More expensive

3.2 Physical method

Advantages

- ✓ Avoids the incorporation of impurities coming from the activating agent
- ✓ The process is not corrosive
- ✓ A washing stage is not required
- ✓ Cheaper

Disadvantages

- ✓ The activated carbons are obtained in two steps
- ✓ Higher temperatures of activation (800-1000 ºC)
- ✓ Poorer control of the porosity

4. Conclusions

Attending to the works considered in this chapter, chemical activation is the most used method for the preparation of ACs (~60 %) from lignocellulosic precursors. Physical activation methods is used in 28% of the studies and a low quantity of studies combine both methods (i.e., physicochemical) to produce ACs. H_3PO_4 and $ZnCl_2$ are the two more employed activating agents in the impregnation of lignocellulosic materials (30% and 24 %, respectively), whereas alkaline reagents such as KOH, NaOH and K_2CO_3 have been considered because ACs with high specific surface can be obtained (1500-2500 $m^2\ g^{-1}$). Physical activation of lignocellulosic precursors normally renders carbons with lower specific surface area. However, when compared with chemical activation, this method is not corrosive and does not require a washing step.

5. Acknowledgments

The author thanks the support of CONACYT (AGS-2010-C02-143917), DGEST (4220.11-P), Instituto Tecnológico de Aguascalientes (México) and Instituto Nacional del Carbón (Oviedo, España).

6. References

[1] Adinata, D., Wan-Daud, W.M. & Kheireddine-Aroua, M. (2007). Preparation and characterization of activated carbon from palm shell by chemical activation with K_2CO_3. *Bioresource Technology,* Vol. 98, No. 1, (January 2007), pp. (145–149), ISSN 0960-8524.

[2] Ahmedna, M., Marshall, W.E. & Rao, R.M. (2000). Production of granular activated carbons from select agricultural by-products and evaluation of their physical, chemical and adsorption properties. *Bioresource Technology,* Vol. 71, No. 2, (January 2000), pp. (113-1239, ISSN 0960-8524. Arami-Niya, A., Daud, W.M.A.W. & Mjalli, F.S. (2010). Using granular activated carbon prepared from oil palm shell by $ZnCl_2$ and physical activation for methane adsorption. *Journal of Analytical and Applied Pyrolysis,* Vol. 89, No. 2, (November 2010), pp. (197-203), ISSN 0165-2370.

[3] Arami-Niya, A., Daud, W.M.A.W. & Mjalli, F.S. (2011). Comparative study of the textural characteristics of oil palm shell activated carbon produced by chemical and physical activation for methane adsorption. *Chemical Engineering Research and Design,* Vol. 89, No. 6, (June 2011), pp. (657–664), ISSN 0263-8762.

[4] Aravindhan, R., Raghava-Rao, J. & Unni-Nair, B. (2009). Preparation and characterization of activated carbon from marine macro-algal biomass. *Journal of Hazardous Materials,* Vol. 162, No. 2-3, (March 2009), pp. (688–694), ISSN 0304-3894.

[5] Aworn, A., Thiravetyan, P. & Nakbanpote, W. (2008). Preparation and characteristics of agricultural waste activated carbon by physical activation having micro and mesopores. *Journal of Analytical and Applied Pyrolysis,* Vol. 82, No. 2, (July 2008), pp. (279–285), ISSN 0165-2370.

[6] Azevedo, D.C.S., Araújo, J.C.S., Bastos-Neto, M., Torres, A.E.B., Jaguaribe, E.E. & Cavalcante C.L. (2007). Microporous activated carbon prepared from coconut shells using chemical activation with zinc chloride. *Microporous and Mesoporous Materials,* Vol. 100, No. 1-3, (March 2007), pp. (361-364), ISSN 1387-1811.

[7] Bagheri, N. & Abedi, J. (2009). Preparation of high surface area activated carbon from corn by chemical activation using potassium hydroxide. *Chemical Engineering Research and Design,* Vol. 87, No. 8, (August 2009), pp. (1059–1064), ISSN 0263-8762.

[8] Boudrahem, F., Aissani-Benissad, F. & Aït Amar, H. (2009). Batch sorption dynamics and equilibrium for the removal of lead ions from aqueous phase using activated carbon developed from coffee residue activated with zinc chloride. *Journal of Environmental Management,* Vol. 90, No. 10, (July 2009), pp. (3031–3039), ISSN 0301-4797.

[9] Carvalho, A., Gomes, M., Mestre, A.S., Pires, J. & Brotas de Carvalho, M. (2004). Activated carbons from cork waste by chemical activation with K_2CO_3. Application to adsorption of natural gas components. *Carbon,* Vol. 42, No. 3, (January 2004), pp. (667-691), ISSN 0008-6223.

[10] Caturla, F., Molina-Sabio, M., & Rodríguez-Reynoso, F. (1991). Preparation of activated carbon by Chemical activation with $ZnCl_2$. *Carbon,* Vol. 29, No. 7, (February 1991), pp. (999-1007), ISSN 0008-6223.

[11] Elizalde-González, M.P., Mattusch, J., Peláez-Cid, A.A. & Wennrich, R. (2007). Characterization of adsorbent materials preparaed from avocado kernel sedes: Natural, activated and carbonized forms. *Journal of Analytical and Applied Pyrolysis,* Vol. 78, No. 1, pp. (185-193), ISSN 0165-2370.

[12] Erdoğan, S., Önal, Y., Akmil-Başar, C., Bilmez-Erdemoğlu, S., Sarıcı-Özdemir, Ç., Köseoğlu, E. & İçduygu, G. (2005). Optimization of nickel adsorption from aqueous solution by using activated carbon prepared from waste apricot by chemical activation. *Applied Surface Science*, Vol. 252, No. 5, (December 2005), pp. (1324–1331), ISSN 0169-4332.

[13] Foo, K.Y. & Hameed, B.H. (2011). Preparation and characterization of activated carbon from pistachio nut shells via microwave-induced chemical activation. *Biomass and Energy*, Vol. 35, No. 7, (July 2011), pp. (3257-3261), ISSN 0961-9534.

[14] González, J.F., Román, S., Encinar, J.M. & Martínez, G. (2009). Pyrolysis of various biomass residues and char utilization for the production of activated carbons. *Journal of Analytical and Applied Pyrolysis*, Vol. 85, No. 1-2, (MAy 2009), pp. (134–141), ISSN 0165-2370.

[15] Hared, I.A., Dirion, J.L., Salvador, S., Lacroix, M. & Rio, S. (2007). Pyrolysis of wood impregnated with phosphoric acid for the production of activated carbon: Kinetics and porosity development studies. *Journal of Analytical and Applied Pyrolysis*, Vol. 79, No. 1-2, (May 2007), pp. (101–105), ISSN 0165-2370.

[16] Hayashi, J., Horikawa, T., Takeda, I., Muroyama, K. & Ani, F.N. (2002a). Preparing activated carbon from various nutshells by chemical activation with K_2CO_3. *Carbon*, Vol. 40, No. 13, (April 2002), pp. (2381–2386), ISSN 0008-6223.

[17] Hayashi, J., Horikawa, T., Takeda, I., Muroyama, K. & Ani, F.N. (2002b). Activated carbon from chickpea husk by chemical activation with K_2CO_3: preparation and characterization. *Microporous and Mesoporous Materials*, Vol. 55, No. 1, (August 2002), pp. (63-68), ISSN 1387-1811.

[18] Hazourli S., Ziati M. & Hazourli A. (2009). Characterization of activated carbon prepared from lignocellulosic natural residue:-Example of date stones-. *Physics Procedia*, Vol. 2, No. 3, (November 2009), pp. 1039-1043, ISSN 1875-3892.

[19] Hu, Z., Srinivasan, M.P. & Ni, Y. (2001). Novel activation process for preparing highly microporous and mesoporous activated carbons. *Carbon*, Vol. 39, No. 6, (May 2001), pp. (877-886), ISSN 0008-6223.

[20] Jibril, B., Houache, O., Al-Maamari, R. & Al-Rashidi, B. (2008). Effects of H_3PO_4 and KOH in carbonization of lignocellulosic material. *Journal of Analytical and Applied Pyrolysis*, Vol. 83, No. 2, (November 2008), pp. (151-156), ISSN 0165-2370.

[21] Juárez-Galán, J., Silvestre-Albero, A., Silvestre-Albero, J. & Rodríguez-Reinoso, F. (2009). Synthesis of activated carbon with highly developed "mesoporosity". *Microporous and Mesoporous Materials*, Vol. 117, No. 1-2, (January 2009) pp. 519-521, ISSN 1387-1811.

[22] Kaghazchi, T., Asasian-Kolur, N. & Soleimani, M. (2010). Licorice residue and Pistachio-nut shell mixture: A promising precursor for activated carbon. *Journal of Industrial and Engineering Chemistry*, Vol. 16, No. 3, (May 2010), pp. (368–374), ISSN 1226-086X.

[23] Lua, A.C. & Yang, T. (2004). Effect of activation temperature on the textural and chemical properties of potassium hydroxide activated carbon prepared from pistachio-nut shell. *Journal of Colloid and Interface Science*, Vol. 274, No. 2, (June 2004), pp. (594–601), ISSN 0021-9797.

[24] Lua, A.C., Lau, F.Y. & Guo, J. (2006). Influence of pyrolysis conditions on pore development of oil-palm-shell activated carbons. *Journal of Analytical and Applied Pyrolysis*, Vol. 76, No. 1-2, (June 2006), pp. (96–102), ISSN 0165-2370.

[25] Mohamed, A. R., Mohammadi, M. & Darzi, G.N. (2010). Preparation of carbon molecular sieve from lignocellulosic biomass: A review. *Renewable and Sustainable Energy Reviews,* Vol. 14, No. 6, (August 2010), pp. (1591-1599), ISSN 1364-0321.
[26] Mohammadi, S.Z., Karimi, M.A., Afzali, D. & Mansouri, F. (2010). Removal of Pb(II) from aqueous solutions using activated carbon from Sea-buckthorn stones by chemical activation. *Desalination,* Vol. 262, No. 1-3, (November 2010), pp. (86–93), ISSN 0011-9164.
[27] Moreno-Castilla, C., Carrasco-Marín, F., López-Ramón, M.V. & Alvarez-Merino, M.A. (2001). Chemical and physical activation of olive-mill waste water to produce activated carbons. *Carbon,* Vol. 39, No. 9, (August 2001), pp. (1415–1420), ISSN 0008-6223.
[28] Mourão, P.A.M., Laginhas, C., Custódio, F., Nabais, J.M.V., Carrott, M.M.L. & Ribeiro-Carrot, M.M.L. (2011). Influence of oxidation process on the adsorption capacity of activated carbons from lignocellulosic precursors. *Fuel Processing Technology,* Vol. 92, No. 2, (February 2011), pp. (241-246), ISSN 0378-3820.
[29] Nabais, J.V., Carrott, P., Ribeiro-Carrott, M.M.L., Luz, V. & Ortiz, A.L. (2008). Influence of preparation conditions in the textural and chemical properties of activated carbons from a novel biomass precursor: The coffee endocarp. *Bioresource Technology,* Vol. 99, No. 15, (October 2008), pp. (7224–7231), ISSN 0960-8524.
[30] Namasivayam, C. & Kadirvelu, K. (1997). Activated carbons prepared from coir pith by Physical and Chemical activation methods. *Bioresource Technology,* Vol. 62, No. 3, (December 1997), pp. (123-127), ISSN 0960-8524.
[31] Olivares-Marín, M., Fernández-González, C., Macías-García, A. & Gómez-Serrano, V. (2006). Preparation of activated carbon from cherry stones by chemical activation with $ZnCl_2$. *Applied Surface Science,* Vol. 252, No. 17, (June 2006), pp. (5967–5971), ISSN 0169-4332.
[32] Phan, N.H., Rio, S., Faur, C., Le Coq, L., Le Cloirec, P. & Nguyen, T.H. (2006). Production of fibrous activated carbons from natural cellulose (jute, coconut) fibers for water treatment applications. *Carbon,* Vol. 44, No. 12, (October 2006), pp. (2569–2577, ISSN 0008-6223).
[33] Prahas, D., Kartika, Y., Indraswati, N. & Ismadji, S. (2008). Activated carbon from jackfruit peel waste by H_3PO_4 chemical activation: Pore structure and surface chemistry characterization. *Chemical Engineering Journal,* Vol. 140, No. 1-3, (July 2008), pp. (32-42), ISSN 1385-8947.
[34] Puziy, A.M., Poddubnaya, O.I., Martínez-Alonso, A., Suárez-García, F. & Tascón, J. (2005). Surface chemistry of phosphorus-containing carbons of lignocellulosic origin. *Carbon,* Vol. 43, No. 14, (November 2005), pp. (2857-2868), ISSN 0008-6223.
[35] Rodriguez-Mirasol, J., Cordero, T. & Rodríguez, J.J. (1993). Preparation and characterization of activated Carbons from eucalyptus krafl- lignin. *Carbon,* Vol. 31, No. 1, (May 1992), pp. (87-95), ISSN 0008-6223.
[36] Rodríguez-Reinoso, F. & Molina-Sabio, M. (1992). Activated carbons from lignocellulosic materials by chemical and/or physical activation: an overview. *Carbon,* Vol. 30, No. 7, (1992), pp. (1111-1118), ISSN 0008-6223
[37] Sudaryanto, Y., Hartono, S.B., Irawaty, W., Hindarso, H. & Ismadji, S. (2006). High surface area activated carbon prepared from cassava peel by chemical activation. *Bioresource Technology,* Vol. 97, No. 5, (March 2006), pp. (734–739), ISSN 0960-8524.

[38] Tay, T., Ucar, S. & Karagöz, S. (2009). Preparation and characterization of activated carbon from waste biomass. *Journal of Hazardous Materials,* Vol. 165, No. 1-3, (June 2009), pp. (481–485), ISSN 0304-3894.

[39] Tongpoothorn, W., Sriuttha, M., Homchan, P., Chanthai, S. & Ruangviriyachai, C. (2011). Preparation of activated carbon derived from Jatropha curcas fruit shell by simple thermo-chemical activation and characterization of their physico-chemical properties *Chemical Engineering Research and Design,* Vol. 89, No. 3, (March 2011), pp. (335–340), ISSN 0263-8762.

[40] Torregrosa, R. & Martín-Martínez, J.M. (1991). Activation of lignocellulosic materials: a comparison between chemoical, physical and combined activation in terms of porous texture. *Fuel,* Vol. 70, No. 10, (October 1991), pp. (1173-1180), ISSN 0016-2361.

[41] Tseng, R. (2007). Physical and chemical properties and adsorption type of activated carbon prepared from plum kernels by NaOH activation. *Journal of Hazardous Materials,* Vol. 147, No. 3, (August 2007), pp. (1020–1027), ISSN 0304-3894.

[42] Valix, M., Cheung, W.H. & McKay, G. (2004). Preparation of activated carbon using low temperature carbonisation and physical activation of high ash raw bagasse for acid dye adsorption. *Chemosphere,* Vol. 56, No. 5, (August 2004) pp. (493–501), ISSN 0045-6535.

[43] Vargas, J.E., Gutierrez, L.G. & Moreno-Piraján, J.C. (2010). Preparation of activated carbons from seeds of Mucuna mutisiana by physical activation with steam. *Journal of Analytical and Applied Pyrolysis,* Vol. 89, No. 2, (April 2001), pp. (307–312), ISSN 0165-2370.

[44] Vargas, A.M.M., Cazetta, A.L., Garcia, C.A., Moraes, J.C.G., Nogami, E.M., Lenzi, E., Costa W.F. & Almeida, V.C. (2011). Preparation and characterization of activated carbon from a new raw lignocellulosic material: Flamboyant (Delonix regia) pods. *Journal of Environmental Management,* Vol. 92, No. 1, (January 2011), pp. (178-184), ISSN 0301-4797.

[45] Warhurst, A.M., Fowler, G.D., McConnachie, G.L. & Pollard, S.J.T. (1997). Pore structure and adsorption characteristics of steam pyrolysis carbons from Moringa Oleifera. *Carbon,* Vol. 35, No. 8, (February 1997), pp. (1039-1045), ISSN 0008-6223.

[46] Yagmur, E., Ozmak, M. & Aktas, Z. (2008). A novel method for production of activated carbon from waste tea by chemical activation with microwave energy. *Fuel,* Vol. 87, No. 15-16, (November 2008), pp. (3278–3285), ISSN 0016-2361.

[47] Yang, J. & Qiu, K. (2010). Preparation of activated carbons from walnut shells via vacuum chemical activation and their application for methylene blue removal. *Chemical Engineering Journal,* Vol. 165, No. 1, (November 2010), pp. (209-217), ISSN 1385-8947.

[48] Yavuz, R., Akyildiz, H., Karatepe N. & Çetinkaya, E. (2010). Influence of preparation conditions on porous structures of olive stone activated by H_3PO_4. *Fuel Processing Technology,* Vol. 91, No. 1, (January 2010), pp. (80-87), ISSN 0378-3820.

[49] Zuo, S., Yang, J., Liu, J. & Cai, X. (2009). Significance of the carbonization of volatile pyrolytic products on the properties of activated carbons from phosphoric acid activation of lignocellulosic material. *Fuel Processing Technology,* Vol. 90, No. 7-8, (July-August 2009), pp. (994-1001), ISSN 0378-3820.

3

Techniques Employed in the Physicochemical Characterization of Activated Carbons

Carlos J. Durán-Valle
Universidad de Extremadura
Spain

1. Introduction

Activated carbons have been widely used as adsorbents, catalyst supports, catalysts, and electronic materials due to its properties: high surface area, large pore volume, and chemical-modifiable surface (Do, 1998). These properties determine its application.

Porous solids are materials full of pores; when the main type of pore is microporous, these materials have large internal surfaces. This property is of great importance in adsorption, and its study is essential in the characterization of activated carbons. Pore structure and specific surface can be controlled by several factors: carbonization atmosphere, activation agent, precursor, time and temperature of thermal treatment, the use of templates to synthesize the precursor, particle size, and chemical treatment. The same factors, but with different intensity, control the functional groups in the surface. Also, the functional groups in activated carbons were found to be responsible for the variety in physical and chemical properties (Bandosz, 2009). So, a great amount research has focused on how to modify and characterize the surface functional groups of activated carbons in order to improve their applications or understand their properties (Calvino-Casilda et al., 2010; Moreno-Castilla, 2004; Shen et al., 2008). In the last decades, a large number of techniques have been developed to study various surface properties in solids (Somorjai, 1994). The frontiers of instrumentation are constantly being pushed toward better conditions: finer detail (spatial resolution, energy resolution, and composition accuracy), better sensibility, automation, and cheaper equipment. Activated carbon presents some peculiarities: it is a non-crystalline material (amorphous), is a non-stoichiometric solid with variable composition, and it is opaque to most wavelengths and species used in a characterization laboratory. The chemistry of carbon is extremely complex (Schlogl, 1997; Do, 1998); the main reason for this complexity lies in the pronounced tendency of this element to form homonuclear bonds in three bonding geometries (*sp*, sp^2, and sp^3) and in the moderate electronegativity of carbon, which allows strong covalent interactions with other elements. Because of the use of only one technique does not provide all the necessary information about surfaces, the tendency is to use a combination of techniques. In this chapter, some of the more common techniques used in the characterization of activated carbons are shown.

2. Density measurements and pore volume

There are several "densities" related with activated carbons. These can be mentioned as:

- **bulk density**, which can be defined as the volume of a full recipient that contains a determined mass of activated carbon. This volume includes: a) activated carbon's atoms; b) pores; and c) space between particles.

- **mercury density**, where this liquid is used to refill the space between particles, but its high surface tension maintains Hg out of pores.

- **true density** or **helium density**, Helium, as gas with small size molecules, can go into pores (really, only accessible to pores according to the molecular size of helium). The helium density gives a measure of the density of the carbon structure.

The total volume of pores can be calculated from Hg and He densities by:

$$V = \frac{1}{\rho_{He}} - \frac{1}{\rho_{Hg}} \quad (1)$$

where V is the pore volume, ρ_{He} is the He density, and ρ_{Hg} is Hg density, respectively.

In preparation of charcoal, these values have been studied for different temperatures and times of preparation of charcoal (Durán-Valle et al., 2006; Pastor-Villegas & Durán-Valle, 2002).

3. Porosimetry

The pores of solids are of different kinds. The individual pores may vary in size and in shape. With respect to the shape, in activated carbons, the predominant type is the slit-shape pore. But, the width of the pores is also of special interest for many purposes. A classification of pores according to their average width, which was adopted by the IUPAC (IUPAC, 1972; Sing et al., 1985) is shown in Table 1.

In recent years, the micropore range has been subdivided into very narrow pores (until 0.8 nm) or **ultramicropores**, where the enhancement of interaction potential is caused by the similarity in size between the pore and molecules, and **supermicropores**, which have a width (0.8 to 2.0 nm) between ultramicropores and mesopores (Gregg & Sing, 1982). Carbon-based materials usually have a bimodal pore size distribution, with one dominant peak being less than approximately 2 nm and the other major peak usually greater than 50 nm (Do et al., 2008).

The volume of macropores it is usually on the order of 0.2-0.5 cm^3 g^{-1}, but the associated area is very small, on the order of 0.5 m^2 g^{-1}, which is negligible in an activated carbon (Do, 1998). Macropores are not important for the adsorption capacity, but their importance is because they act as transport pores to the meso- and micropores. Mesopores have a volume in the range of 0.1 to 0.4 cm^3 g^{-1}, and the surface area is in the range of 10-100 m^2 g^{-1}. Their contribution to adsorption is significant, and they act as transport to micropores. Micropores have a similar volume, but the surface area is the most important, sometimes near to 1000 m^2 g^{-1}.

Type	Width (nm)
Micropores	< 2
Mesopores	>2 and < 50
Macropores	> 50

Table 1. IUPAC classification of pores according to their width.

The characterization of porous activated carbon and its derivatives has been a subject of great interest for many decades. Two main experimental tools are used for this study: mercury porosimetry and gas adsorption.

Mercury porosimetry is a technique that was originally developed to determine pore sizes in the macropore range, where the gas adsorption is not adequate (Gregg & Sing, 1982). Since the contact angle of mercury with solids is >90°, an excess pressure p is required to force the liquid mercury into a pore of radius r. Washburn (1921) suggested the following basic equation:

$$r = \frac{2\gamma \cos\theta}{p} \tag{2}$$

where γ is the surface tension of mercury (484 mN m^{-1}) and θ is the contact angle (141°) recommended by IUPAC (Sing et al, 1985).

The technique of mercury porosimetry consists of measuring the extent of mercury penetration into the pores of a solid as a function of the applied pressure. Automatic porosimeters are now in use for the routine examination of pore structures for catalysts, adsorbents, cements, refractory materials, and other materials. The measure range of porosimeters application extends from macropores to near of the limit between micropores and mesopores. Thus, there is an overlap with the gas adsorption method in the mesopore range.

4. Gas adsorption

The ability to adsorb and desorb gases from the coals has been known for a long time (Scheele, 1780). A large amount of the information about the physical structure of a surface (i.e., specific surface and pores) comes from the amount of gas adsorbed on this surface as a function of gas pressure at a given temperature. Mercury porosimetry and gas adsorption are complementary techniques: porosimetry can measure mesopores and macropores, and gas adsorption measures micropores and mesopores.

The curves derived from these experiments are called **adsorption isotherms**. They can be used to determine thermodynamic parameters (e.g., heat of adsorption), the pore distribution, and the surface area. The IUPAC recommends a classification based on six types of isotherms these are shown in Figure 1. Types I to V were proposed by Brunauer, Deming, Deming, and Teller (Brunauer et al., 1940) and they are referred to as BDDT. Type VI was subsequently added (Gregg & Sing, 1982). There are a considerable number of borderline cases that are difficult to assign to one group rather than another.

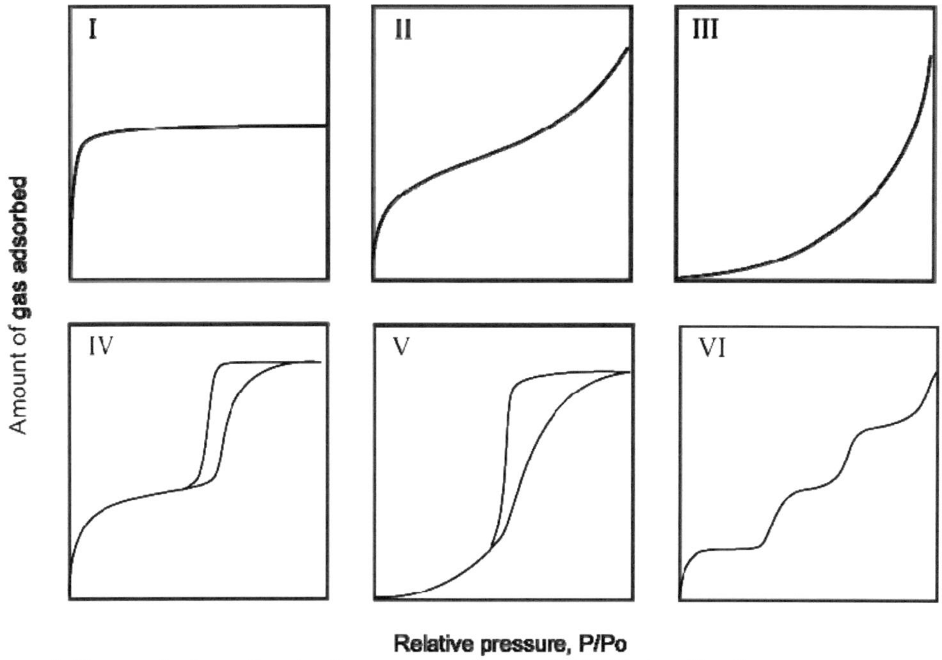

Figure 1. Types of adsorption isotherms according to IUPAC classifications.

Type I isotherms are characteristics of the existence of only strong interactions between the adsorbate and adsorbent, which explains the high adsorption at low relative pressure. In activated carbons, this is due to the existence of micropores. If the isotherm is clearly of type I, the carbon is called "microporous carbon," and it can be assumed that most of the porosity is formed by pores whose widths less than 2 nm. This isotherm is very common in activated carbons. The isotherms of type II are generally associated to non-porous solids (Gregg & Sing, 1982). But, if the activated carbon has a wide distribution of pore widths (i.e., micropores and mesopores of different widths), the isotherm obtained can be similar to type II but with higher gas adsorption (Yates, 2003). In these cases, it must be thought that this isotherm is a combination of types I and IV. Pure type II represents the multilayer adsorption of a vapor into macropores or external surface. Type IV is an isotherm characteristic of a mesoporous solid and the hysteresis is due to capillary condensation into mesopores. Isotherms types III and V are characteristics for systems where the adsorbent-adsorbate interaction is weak compared with adsorbate-adsorbate interactions. As activated carbons are a universal adsorbent and the interactions with adsorbate are never weak, then these isotherms are not frequently found in activated carbons. The same condition applies to type VI which is of theoretical interest but relatively rare in activated carbons.

There are a lot of theoretical models applied on gas isotherms data, which allow obtaining physical characterization of carbon surfaces (Do et al, 2008). Some of the most used are cited in Table 2.

Model	Determined parameter	Reference
Barret, Joyner and Halenda (BJH)	Mesopore size distribution	Barret et al. (1951)
Dubinin-Radushkevich (DR) and Dubinin-Astakhov (DA)	Micropore volume	Dubinin et al. (1947)
Horvarth-Kawazoe (HK)	Micropore size distribution	Hovarth & Kawazoe (1983)
Density Functional Theory (NLDFT or DFT)	Pore size distribution	Olivier (1995)
Brunauer, Emmett and Teller (BET)	Surface area	Brunauer et al. (1938)
Langmuir	Surface area	Langmuir (1918)

Table 2. Most common isotherm models used in gas adsorption.

One of the most used is that proposed by Brunauer, Emmett, and Teller: the BET model (Brunauer et al., 1938). This model is a two-parameter adsorption equation of the form (Gregg & Sing, 1982):

$$\frac{P}{\sigma(P_0 - P)} = \frac{1}{\sigma_0 c} + \frac{c-1}{\sigma_0 c}\frac{P}{P_0} \tag{3}$$

where P_0 is the saturation pressure of vapor used, σ is the adsorbed quantity at pressure P, σ_0 is the monolayer capacity, and c is a constant (at isotherm temperature) that depends on the heat of condensation the heat of adsorption. A plot of $P/[\sigma(P_0-P)]$ versus P/P_0 yields a (near) straight line with slope $(c-1)/\sigma_0 c$ and intercept $1/\sigma_0 c$. For type II and IV isotherms, the recommended pressure range is between 0.05 and 0.35 P/P_0 and from 0.02 to 0.12 for type I. If the molecular area occupied by the adsorbed gas is known, the surface area can be calculated from σ_0. The value of c gives information about the adsorbate-adsorbent interaction. When c increases, the interaction is stronger, relative to interactions between molecules of adsorbate, or, in other words, the isotherm type changes from type III (low c) to type II (medium c) and, finally, type I (high value of c). Care should be taken in analysing data, since type III and type V isotherms are not adequate for this mathematical approach.

The most used gas is N_2 (molecular area, 0.162 nm^2) at 77 K, but other vapors have been employed from time to time. The different molecular sizes (and shapes) cause that not all gases can access at the same pores. It is evident that this procedure must lead to anomalous results for the area of a given solid if different gases are used. Prior to the determination of an isotherm, all physisorbed material must be removed from the surface. This is achieved by exposure of the solid to high vacuum and heat. The exact temperature and residual pressure can condition the final results.

5. Proximate analysis

Proximate analysis is one of the thermal analysis techniques. These are analytical techniques in which a physical property is measured with a temperature-programmed variation. In this case, the property measured is the weight.

Proximate analysis provides an approach to estimate the content of: a) moisture; b) volatile matter; c) fixed carbon; and d) ash. There are some standard methods (for example, ASTM, DIN, UNE), but the analysis is usually carried out in an automated thermogravimetric system. Moisture measurements refer to the matter volatilized until near 373 K in an inert atmosphere, mostly water. Volatile matter is determined by the same procedure but in a temperature range of 373 to 1223 K. Fixed carbon is the material burned in air at 1223 K in a third step and is made of the more stable organic structures. Lastly, the non-combustible matter is the ash. Generally, the composition of ash, fixed carbon, and volatile matter are given on a dry basis. The representation of weight versus temperature (or time) also gives information about the thermal stability of the activated carbon.

The proximate analysis is susceptible of variations in the thermal treatment. Generally, when temperature or time increases, the volatile matter content decreases. The effect of this is that both the ash content and the fixed carbon content increase as a consequence of the concentration of the inorganic fraction and carbon in charcoal (Duran-Valle et al., 2006).

As in the elemental analysis, this technique does not give accurate information about functional groups but can provide limited information on chemical structure (see Table 3). A better and related technique is thermal programmed desorption, in which desorbed species produced at heating are analyzed (see below).

Element	Low content	High content
Moisture	High HHV Low graphitization grade	Low HHV
Volatile matter	High graphitization grade Low amount of functional groups	High graphitization grade High HHV High amount of functional groups
Fixed carbon	Low graphitization grade High amount of functional groups	High graphitization grade Low amount of functional groups
Ash	---	Low HHV

Table 3. Information about the chemistry of carbon obtained from proximate analysis.

6. Elemental analysis

Elemental analysis is the primary method to obtain knowledge about carbon chemistry (Bandosz, 2009). This technique does not provide details on functional groups but at least gives information about heteroatom content and it can provide approximate information on chemical structure, see Table 3 (Chingombe et al., 2005), graphene size (Duran-Valle, 2006), and pore structure (Pastor-Villegas et al., 1998). There are two types of elemental analysis: organic and inorganic. The organic elemental analysis is accomplished by combustion analysis and generally determines carbon, hydrogen, nitrogen, sulphur, and, by difference, oxygen; observations about each one are reported in Table 4.

Element	Low content	High content
C	-Low higher heating value (HHV) -Low graphitization grade	-High HHV -High graphitization grade -Low amount of functional groups -Probably non-polar surface (with low content of O and N) -Probably basic surface (with low content of O)
H	-High graphitization grade	-Low graphitization grade
O	-Low amount of oxygenated functional groups	-High amount of functional groups -Polar surface -Acidic surface
N	---	-Polar surface -Basic surface

Table 4. Information about the chemistry of carbon obtained from organic elemental analysis.

The inorganic elemental analysis yields information about inorganic material (ashes in carbon materials, supported catalysts) and can be carried out by various techniques, based mainly on X-ray, electron, or mass spectroscopy. Related with elemental analysis is radioactive characterization (Rubio-Montero et al., 2009), this is a promising method to differentiate between charcoals (carbonized biomass) and coals.

7. Acid/base titration

One of the most influential variables in the adsorption in solution is the pH, since this parameter can change the sign (or presence) of charges onto the adsorbate. On the surface of activated carbon, the predominant charge (positive or negative) depends on the acidic or basic character of the adsorbent. Therefore, it is important to study the presence of acidic or basic functional groups on activated carbons.

A global measurement of the acidity/basicity of a carbon is the point of zero charge (PZC) at which the surface charge density is 0 (IUPAC, 1997). It is usually determined in relation to a disolution's pH: the pH of the solution at equilibrium with a solid when the solid exhibits zero net electrical charge on the surface. A very similar concept is the isoelectric point (IEP), which requires that the charge is 0 in the entire solid, not only on the surface. Generally, IEP is very similar to PZC. The PZC can be obtained from acid-base titrations of dispersions and monitoring the electrophoretic mobility of the particles and the solution pH. The titration of several dispersions of activated carbons in aqueous solutions with different initial pHs is easier and cheaper. A plot of final pH versus initial pH of the solutions with and without activated carbon gives a cross that indicates the PZC (see Figure 2).

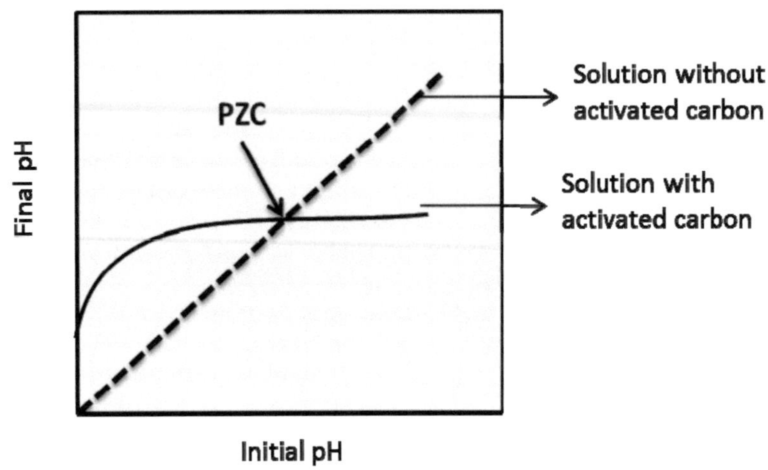

Figure 2. PZC measurement determined in several solutions at different initial pHs.

Another method is also mass titration with only a solution (whose pH in this method is irrelevant) and with at least 7% (w/v) activated carbon (Valente-Nabais & Carrot, 2006). The pH of the solution is approaching to the PZC when the carbon quantity increases. With 7% (w/v) of activated carbon, the equilibrium is reached. Then, the solution pH is the PZC of the adsorbent (see Figure 3).

A more detailed assessment of the acidity/basicity is performed by titration with several substances of different acidities. As an example, in the method proposed by Boehm (Boehm, 2002), the amount of oxygen-containing groups (carboxyl, lactonic, phenol, and others) was determined by adsorption neutralization with $NaHCO_3$, Na_2CO_3, NaOH, and $NaOCH_2CH_3$ solutions, respectively. Also, the basic group content can be determined with HCl solution. This method, widely used, has several drawbacks. As in all classic titrations with solids (especially with microporous solids), the equilibrium time is very long. There may be functional groups with the same structure and very different acidities, or different functional groups with similar acidities (for example, 2,4-dinitrophenol and benzoic acid have a similar pK_a). This prevents quantification of functional groups in its structure, although it could be realized by acid strength. Another failure is that functional groups containing other heteroatoms are considered as oxygen functional groups. The acidic or basic functional groups can be also characterized by immersion calorimetry (López-Ramón et al., 1999).

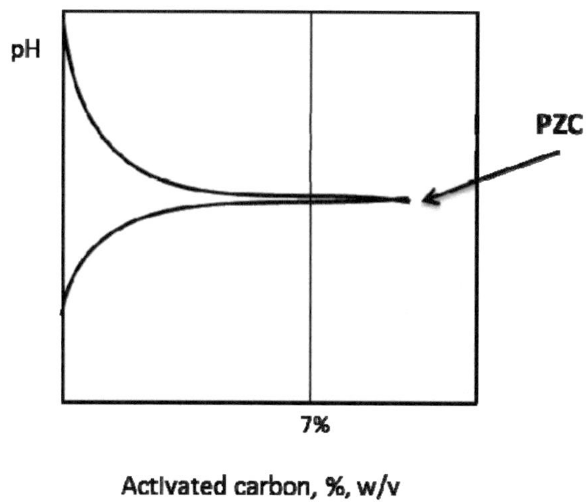

Figure 3. Determination of PZC by Valente-Nabais & Carrott method. Variation of activated carbon w/v.

8. Thermal Programmed Desorption (TPD)

TPD is a thermal analysis method widely used for the characterization of activated carbons. In this technique, a sample is heated in a carrier gas, to induce thermal desorption of adsorbed species or thermal decomposition. The desorbed products are analyzed by several methods. By heating, the oxygenated groups are thermally decomposed, releasing CO, CO_2, and H_2O at different temperatures. The groups can be identified by decomposition temperature and type of gas and can be quantified by the areas of peaks. The major difficulty is the identification of each surface group, because in activated carbons TPD spectra show composite and broad peaks of all gases released. An application of this technique with several references is shown in (Figueiredo et al., 1999).

9. Calorimetric Techniques

There are a set of techniques that measure the heat involved in a process (Bandosz, 2009). Such techniques have been used to estimate some aspects of the physical and the chemical structure of activated carbons. Immersion calorimetry provides a measurement of the energy involved in the interaction of molecules of a liquid with the surface of activated carbon. The use of several liquids with different polarities can be related with the hydrophobic and hydrophilic nature of the surface. The enthalpy of immersion of carbonaceous materials into water can give information about primary adsorption centers that are related to the oxygen content (Stoeckli et al., 1983). The differential scanning calorimetry (DSC) measures the difference in the amount of heat required to increase the temperature of a sample and a reference sample as a function of temperature. A similar technique is the differential thermal analysis; in this, the difference in temperature between sample and reference is the parameter measured. Both techniques can be used to evaluate the thermal behavior of a charcoal (Duran-Valle et al., 2005).

10. Infrared Spectroscopy (FTIR)

Infrared spectroscopy is a traditional method for structural analyses of organic compounds, where infrared radiation is absorbed selectively by the various bonds within a compound. Since FTIR spectroscopy can detect specific bonds in a material, then it is possible to know which functionalities exist on the surface of carbon. IR spectroscopy has been used to study, for example, the changes in the surface chemical structure of the carbon materials after oxidation (Chingombe et al., 2005; Moreno-Castilla et al., 2000; Pradhan & Sandle, 1999), reactions with alkali (Lillo-Ródenas et al., 2003), the carbonization and activation process (Pastor-Villegas et al., 1999), or in the chemical activation of wood (Solum et al., 1995). FTIR is mainly used as a qualitative technique for the analysis of the chemical structure of activated carbons and, sometimes, as quantitative technique (Bandosz, 2009). Two problems are associated with this technique—a) the opacity of carbons and b) the broad peaks—because they are usually a sum of interactions of similar types of functional groups. This technique was used intensively on carbonaceous materials when equipments with Fourier transform (FT) were accessible. The FT allows an improvement over signal/noise rate, energy throughout, accuracy, and fast scans. This approach partially eliminated the opacity problem. Better results can be obtained with alternative techniques (López & Márquez, 2003) that allow reflection (rather than transmission) on activated carbon surfaces—for example, specular reflectance, diffuse reflectance (DRIFT), attenuated total reflectance (ATR), or photoacustic spectroscopy (FTIR-PAS). The assignment of the IR bands to different functional groups is made by comparison with adsorption/transmission bands of organic compounds (Bellamy, 1986; Smith, 1999). The width of the bands is due to the existence of several similar bands with a maximum at different frequencies, because they are affected by vicinal functionalities. Table 5 shows the approximate assignment of main bands in charcoals IR spectra.

Wavenumber (cm^{-1})	Assignment	Structures
3600–3000	Stretching O-H, N-H	Hydroxyl, carboxilic acid
3000–2800	Stretching C-H	Alifatic, olefinic and aromatic hidrocarbons
1770–1650	Stretching C=O	Carbonyl
1700–1600	Stretching C=C	Olefinic structures
1650–1500	Stretching C=C	Aromatic structures
1480–1420	Bending C-H	Alifatic structures
1430–1360	Bending O-H and C-H	Hydroxyl, carboxilic acid, olefines, methyl
1300–1200	Stretching C-O	Unsaturated ethers
1160–1050	Stretching C-O	Tertiary hydroxyl
1120–1070	Stretching C-O	Secondary hydroxyl
1060–1000	Stretching C-O	Primary hydroxyl
900–700	Bending out of the plane C-H	Aromatic structures

Table 5. Assignment of bands in IR spectra of charcoals.

11. X-Ray Photoelectron Spectroscopy (XPS)

XPS (also known as ESCA, electron spectroscopy for chemical analysis) is a technique frequently used in surface chemistry. XPS measures the energy of internal atomic orbitals. This value is characteristic of each element, and so, XPS gives information about elemental composition. But, this value changes slightly with the electric charge on the atom. Therefore, XPS also gives information about functional groups in activated carbons. An important feature of this technique is that the analysis is limited to the surface (some nanometers). This feature can be positive (for surface chemistry studies) or negative (if surface contamination exits). This fact is typical of a spectroscopic technique that uses electrons in their mechanism, because the electrons interact with materials more than photons of the same energy, and so, electrons can pass through a smaller amount of material (in XPS, 0.4 - 4 nm, depending on the kinetic energy of the electron).

XPS uses monochromatic X-ray photons to excite an inner-shell electron. This electron can be extracted out of the atom, and its kinetic energy can be measured. The kinetic energy depends on the binding energy. The mechanism of the process is shown in Figure 4. An incident photon (given is clear color) of energy $h\nu$ is absorbed by an atom, and an electron (given in dark color) placed in an orbital of bind energy E_b leaves the atom with a kinetic energy E_e. The energy of the photon is divided between the energy to translate the electron to the Fermi level (E_b), the kinetic energy, and a correction factor, the work function (w), which depends on the equipment. The energy equation is:

$$h\nu = E_b + E_e + w \tag{4}$$

All data except E_b are experimentally known, and E_b can be easily calculated. The intensity of the peaks is related to the concentration of the element, and quantitative analysis (of the surface) can be carried out.

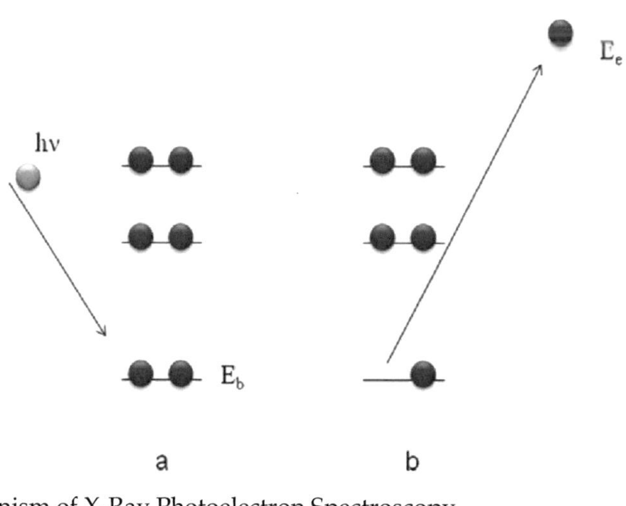

Figure 4. Mechanism of X-Ray Photoelectron Spectroscopy.

In activated carbons, XPS is used to characterize surface functionalities (Bandosz, 2009; Chingombe et al., 2005; Moreno-Castilla et al., 2000) employing binding energies of the C 1s, N 1s, and O 1s photoelectrons (Figueiredo et al., 1999). A comparison with elemental analyses can yield information about different compositions in the surface and into the bulk (Figueiredo et al., 1999).

Table 6 shows an interpretation of peaks of C 1s XPS spectra. Assignment of peaks in N 1s spectra is more difficult due to the large variety of species (with different oxidation states) that nitrogen atoms can form. The value of binding energy varies between 399 eV (reduced form of nitrogen, as amines) to 404 eV (oxidized form of nitrogen, as nitro groups or nitrate). The assignment of peaks for O 1s spectra is easier, because generally only three bands can be distinguished: near 532 eV (double bond C=O), 534 eV (single bond C-O), and 536 eV (water occluded) (Lin et al., 2008).

Two similar techniques are UPS (ultraviolet photoelectron spectroscopy), where photons have energy until 50 eV which allows to be used to study external orbitals (Raymundo-Piñero et al., 2002); and SXPS (soft X-ray photoelectron spectroscopy), which uses photons with energy from 50 to 150 eV.

Binding energy (eV) (approximate)	Assignment
285.0	Aromatics and aliphatics structures. Carbon atoms bond to hydrogen or carbon atoms
286.0	Single C-O bond (alcohol, ether)
287.5	Double C=O bond (carbonyl)
289.0	O-C=O (carboxyl, ester)
290.5	Carbonate, CO_2
291.5	Plasmon

Table 6. Assignment of peaks in C 1s XPS spectra.

12. X-Ray diffraction (XRD)

Activated carbons contain short-range ordered structures, then XRD can provide details about the crystallites or micrographites into the carbon structure, including the disordered and defective features (Burian et al., 2005). This information is obtained from the distribution of scattered radiation using X-rays. The observed diffraction pattern may be converted to structural data, so XRD is used not only to estimate crystallite size or graphitization degree in carbonaceous materials (Biniak et al., 2010; Pradhan & Sandle, 1999) but also used to characterize the inorganic material (Pastor-Villegas et al., 1999). The degree of graphitization is an important parameter, since it reflects the transition extent of carbon material from turbostratic to graphitic structure, and determines some properties of the material (Hussain et al., 2000; Zou et al., 2003).

Generally, the (0 0 2) line is used to determine the interlayer spacing $d_{(0\,0\,2)}$ according to the Bragg equation. A method for calculating the degree of graphitization (g) can be done by using the following equation (Maire & Mering, 1970):

$$g(\%) = \frac{0.3440 - d_{(002)}}{0.3440 - 0.3354} \times 100 \quad (5)$$

where 0.3440 (nm) is the interlayer spacing of the non-graphitized carbon and 0.3354 (nm) is the interlayer spacing of an ideal-type graphite crystallite.

Other method uses the surface under two peaks (Johnson, 1959). The first of these peaks (A) is broad and situated at low 2θ values (maximum near 20 degrees), and it is indicative of the amorphous nature of the material. The second peak (C) is situated near 25 degrees, is higher and narrower, and indicates the presence of crystallinity. The equation used is:

$$g(\%) = \frac{I_C}{I_A} + KI_A \quad (6)$$

where I_C is the integrate of peak C, I_A is the integrate of peak A, and K is a constant.

13. Other X-Ray techniques

There are other techniques of characterization of activated carbons than X-ray radiation, but they are not as commonly used as XRD or XPS. Some of these techniques are described below.

13.1 X-ray fluorescence (XRF)

XRF is the emission of characteristic X-ray radiation from a material that has been excited by bombarding it with X-ray or gamma photons. The term fluorescence is applied when the absorption of radiation of a specific energy results in the emission of radiation of a different (generally lower) energy. This technique is widely used for elemental analysis, but it gives more limited information that obtained in XPS. The mechanism of this phenomenon is shown in Figure 5, which can be considered as a continuation from Figure 4b. Later than when an electron is expulsed of an atom absorbing the energy of a photon (Figure 5a), a hole is created on the lower orbital. An electron of a higher orbital "falls" into the lower orbital (Figure 5b), and energy is released in the form of a photon (Figure 5c). The energy of this photon is equal to the energy difference of the two orbitals involved, and this quantity is characteristic of this element.

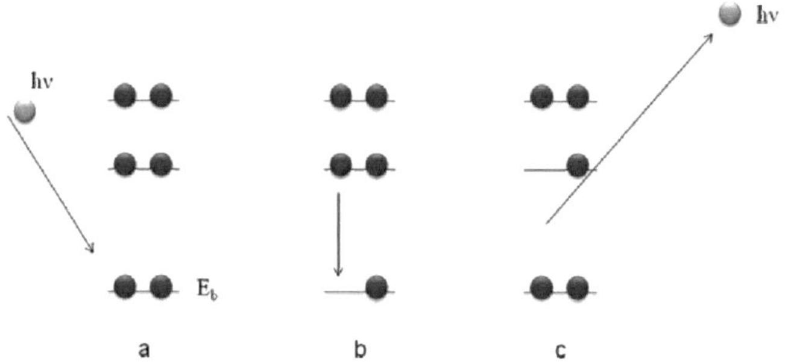

Figure 5. Mechanism of X-ray fluorescence.

13.2 Small-angle X-ray scattering (SAXS)

Small-angle X-ray scattering is a technique where the scattering of X-ray photons by a sample that has inhomogeneities in the nm scale is detected at very low angles. This angular range contains information about pores in activated carbon (Bóta et al., 1997). This technique yields information about physical structure, as XRD, but not about composition, as XPS.

13.3 EDX-SEM

A variation of XRF is the energy-dispersive X-ray spectroscopy (EDX), which is commonly connected to scanning electron microscopes (SEMs). The union of these two techniques allows a point elemental analysis of a surface and also makes a "map" of elemental composition of a surface. This technique is mainly used to study the content and dispersion of metals on a carbon surface.

With illustrative purposes, Figure 6 shows an SEM image (top left) of an activated carbon doped with a sulphur-containing dye. The next figures are the same sample but "tuned" in an element. It can be seen at some points that the concentration of several heteroatoms is high, and these points can be assigned to mineral matter.

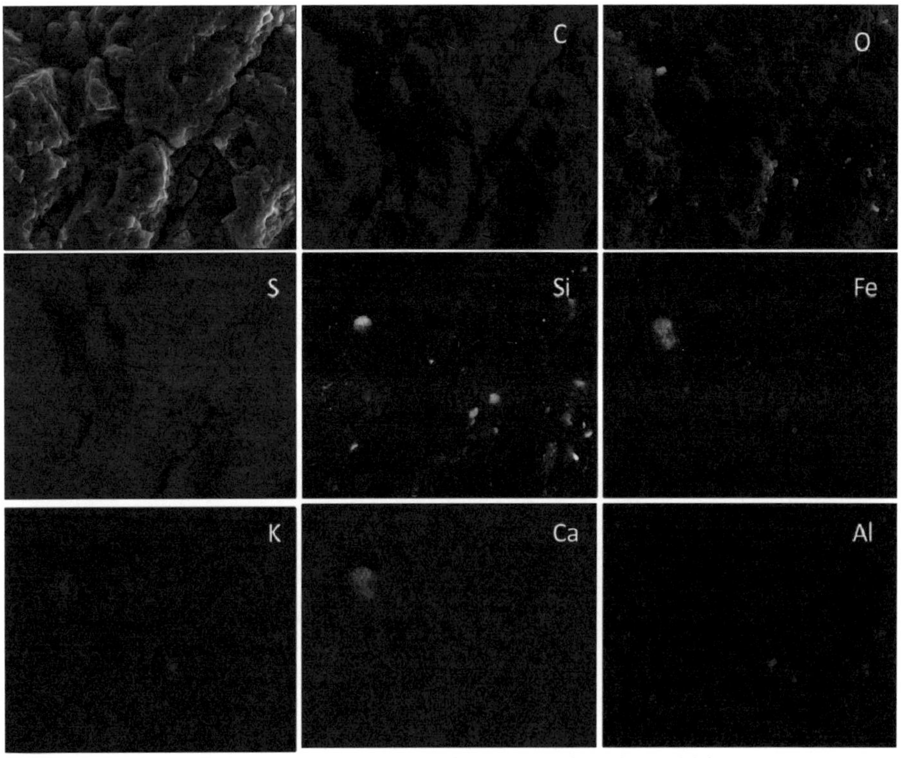

Figure 6. EDX-SEM images of an activated carbon treated with acid blue 25. Image courtesy of Servicio de Apoyo a la Investigación de la Universidad de Extremadura (Badajoz, Spain). Sample supplied by the Instituto Tecnológico de Aguascalientes (Aguascalientes, Mexico).

14. Micro Raman spectroscopy

Raman spectroscopy is used to study vibrational and rotational modes in a system. It is related with FTIR, and both techniques are usually complementary. Laser Raman spectroscopy is used on carbonaceous materials to evaluate the degree of graphitization (Zou et al., 2003) and not for functional groups determination. Micro laser Raman spectroscopy allows to analyze a microscopic surface area of interest. It is an alternative technique to XRD for studying the graphitization extension of a carbonaceous material (Cuesta et al., 1994). In Raman spectra of most activated carbons, two peaks are generally obtained: 1360 and 1580 cm^{-1}. The last one corresponds to graphite structure, and the 1360 cm^{-1} is correlated with a disordered carbon structure. The ratio of the integrated intensities (I_{1360}/I_{1580}) has been considered to be a good parameter to estimate the disorder in the structure.

15. Nuclear Magnetic Resonance (NMR)

NMR is a technique based on measurements of absorption of electromagnetic radiation related with atom nuclei into a magnetic shield. Its use in activated carbon analysis (and other solids) has increased due to the magic-angle spinning (MAS) method that solves the problems of lack of resolution due to the solid state. Generally, the ^{13}C spectrum is the most used. The obtained information includes the hybridization of carbon atoms and the presence of oxygenated functional groups. ^{13}C-NMR can give the aromatic-to-aliphatic carbon ratio and is useful to study changes in surface chemistry (Solum et al., 1995), including heteroatoms. Also, NMR can be used to study compounds adsorbed on carbon surfaces if these compounds have atoms that are different from carbon.

16. Conclusions

Activated carbons have been used for a long time as adsorbents. It is now recognized that activated carbons offer unparallel flexibility in tailoring their physical and chemical properties to specific needs, thus showing the remarkably wide range of potential applications.

The simultaneous use of several physicochemical characterization techniques is very common, and there are a lot of examples in publications. This is due to no one technique may provide all the necessary information about surfaces, together with the extremely complex structures (physical and chemical) of carbon. However, the structure of these materials is not well understood, and more research about the analysis of porous carbonaceous materials is needed.

Traditionally, techniques used in the study of physical structure and those techniques used in the study of chemical structure have been distinguished. But, it has been shown that often both structures are related. It is advisable to conduct a comprehensive study of the activated carbons to better understand their properties.

17. Acknowledgments

The author thanks the support of Spanish Government (CTM2010-14883/TECNO and CTM2010-17776), Junta de Extremadura/FEDER (GRU10011), and Instituto Tecnológico de Aguascalientes (Mexico).

18. References

[1] Bandosz, T.J. (2009). Surface Chemistry of Carbon Materials. In *Carbon Materials for Catalysis*. Serp, P., & Figueiredo, J.L. (Eds.). pp. (58-78), Wiley, ISBN 978-0-470-17885-0.

[2] Barrett, E.P., Joyner, L.G. & Halenda, P.P. (1951). The determination of pore volume and area distributions in porous substances. I. Computations from nitrogen isotherms. *Journal of the American Chemical Society*, Vol. 73, No. 1, (January 1951), pp. (373-380), ISSN 0002-7863.

[3] Bellamy, L.J. (1986). *The infrared spectra of complex molecules*. Chapman and Hall, ISBN 0412138506, USA.

[4] Biniak, S., Pakuła, M., Świątkowski, A., Bystrzejewski, M. & Błażewicz, S. (2010). Influence of high-temperature treatment of granular activated carbon on its structure and electrochemical behavior in aqueous electrolyte solution. *Journal of Material Research*, Vol. 25, No.8, (August 2010), pp. (1617-1628), ISSN 0884-2914.

[5] Boehm, H.P. (2002). Surface oxides on carbon an their analysis : a critical assessment. *Carbon*, Vol. 40, No. 2, (February 2002), pp. (145-149), ISSN 0008-6223.

[6] Bóta, A., László, K., Nagy, L.G. & Copitzky, T. (1997). Comparative study of active carbons from different precursors. *Langmuir*, Vol. 13, No. 24, (November 1997) pp. (6502-6509), ISSN 0743-7463.

[7] Brunauer, S., Deming, L.S.; Deming, W.S. & Teller, E. (1940). On a Theory of the van der Waals Adsorption of Gases. *Journal of the American Chemical Society*, Vol. 62, No. 7, (July 1940), pp. (1723-1732), ISSN 0002-7863.

[8] Brunauer, S., Emmett, P.H. & Teller, E. (1938). Adsorption of Gases in Multimolecular Layers. *Journal of the American Chemical Society*, Vol. 60, No. 2, (February 1938), pp. (309-319), ISSN 0002-7863.

[9] Burian, A., Dore, J.C., Hannon, A.C. & Honkimaki, V. (2005). Influence of high-temperature treatment of granular activated carbon. *Journal of Alloys and Compounds*, Vol. 401, No. 1-2, (September 2005), pp. (18-23), ISSN 0925-8388.

[10] Calvino-Casilda, V., López-Peinado, A.J., Durán-Valle, C.J. & Martín-Aranda, R.M. (2010). Last Decade of Research on Activated Carbons as Catalytic Support in Chemical Processes. *Catalysis Reviews: Science and Engineering*, Vol. 52, No. 3, (2010), pp. (325–380), ISSN 0161-4940.

[11] Chingombe, P., Saha, B. & Wakeman, R.J. (2005). Surface modification and characterisation of a coal-based activated carbon. *Carbon*, Vol. 43, No. 15, (December 2005), pp. (3132-3143), ISSN 0008-6223.

[12] Cuesta, A., Dhamelincourt, P., Laureyns, J., Martínez-Alonso, A. & Tascón, J.M.D., (1994). Raman microprobe studies on carbon materials. *Carbon*, Vol. 32, No. 8, (1994), pp. (1523-1532), ISSN 0008-6223.

[13] Do, D.D. (1998). *Adsorption Analysis: Equilibria and Kinetics*. Imperial College Press, ISBN 1-86094-130-3, London.

[14] Do, D.D.; Ustinov, E.A. & Do, H.D. (2008). Porous Texture Characterization from Gas-Solid Adsorption. In: *Adsorption by Carbons*. Bottani, E.J. & Tascón, J.M.D. (Eds.). pp. (240-270), Elsevier Ltd., ISBN 978-0-08-044464-2, UK.

[15] Dubinin, M.M., Zaverina, E.D. & Radushkevich, L.V. (1947). Sorption and structure of active carbons. I. Adsorption of organic vapors. *Zhurnal Fizicheskoi Khimii*, Vol. 21, (1947), pp. (1351–1362), ISSN 0044-4537.

[16] Durán-Valle, C.J. (2006). Geometrical relationship between elemental composition and molecular size in carbonaceous materials. *Applied Surface Science*, Vol. 252, No. 17, (June 2006), pp. (6097–6101), ISSN 0169-4332.

[17] Durán-Valle, C.J., Gómez-Corzo, M., Gómez-Serrano, V., Pastor-Villegas, J. & Rojas-Cervantes, M.L. (2006). Preparation of charcoal from cherry stones. *Applied Surface Science*, Vol. 252, No. 17, (June 2006), pp. (5957–5960), ISSN 0169-4332.

[18] Durán-Valle, C.J., Gómez-Corzo, M., Pastor-Villegas, J. & Gómez-Serrano, V. (2005). Study of cherry stones as raw material in preparation of carbonaceous adsorbents. *Journal of Analytical and Applied Pyrolysis*, Vol. 73, No. 1, (March 2005), pp. (59-67), ISSN 0165-2370.

[19] Figueiredo, J.L.; Pereira, M.F.R.; Freitas, M.M.A. & Orfao, J.J.M. (1999). Modification of the surface chemistry of activated carbons. *Carbon*, Vol. 37, No. 9, (1999), pp. (1379–1389), ISSN 0008-6223.

[20] Gregg, S.J. & Sing, K.S.W. (1982). *Adsorption, Surface Area and Porosity*. 2nd edition. Academic Press, ISBN 0-12-300956-1, London.

[21] Hovarth, G. & Kawazoe, K. (1983). Method for the calculation of effective pore size distribution in molecular sieve carbon. *Journal of Chemical Engineering of Japan*, Vol. 16, No. 6, (1983), pp. (470-475), ISSN 0021-9592.

[22] Hussain, R., Qadeer, R., Ahmad, M., & Saleem, M. (2000). X-Ray Diffraction Study of Heat-Treated Graphitized and Ungraphitized Carbon. *Turkish Journal of Chemistry*, Vol. 24, No. 2, (June 2000), pp. (177-83), ISSN 1300-0527.

[23] IUPAC (1972). Manual of Symbols and Terminology, appendix 2, Pt.1, Colloid and Surface Chemistry. *Pure and Applied Chemistry*, Vol. 31, No. 4, (1972), pp. (578-638), ISSN1365-3075.

[24] IUPAC, (1997). *Compendium of Chemical Terminology*, 2nd ed. (the "Gold Book"). Compiled by A. D. McNaught & A. Wilkinson. Blackwell Scientific Publications, Oxford (1997). XML on-line corrected version: http://goldbook.iupac.org (2006-) created by M. Nic, J. Jirat, B. Kosata; updates compiled by A. Jenkins. ISBN 0-9678550-9-8.

[25] Johnson, J.E. (1959). X-ray diffraction studies of the crystallinity in polyethylene terephthalate. *Journal of Applied Polymer Science*, Vol. 2, No. 5, (September/October 1959), pp. (205-209), ISSN 0021-8995.

[26] Langmuir I. (1918). The adsorption of gases on plane surfaces of glass, mica and platinum. *Journal of the American Chemical Society*, Vol. 40, No. 9, (September 1918), pp. (1361–1403), ISSN 0002-7863.

[27] Lillo-Ródenas, M.A., Cazorla-Amorós, D. & Linares-Solano, A. (2003). Understanding chemical reactions between carbons and NaOH and KOH: An insight into the chemical activation mechanism. *Carbon*, Vol. 41, No. 2, (February 2003), pp. (267-275), ISSN 0008-6223.

[28] Lin, H.Y., Chen, W.C., Yuan, C.S. & Hung, C.H. (2008). Surface Functional Characteristics (C, O, S) of Waste Tire-Derived Carbon Black before and after Steam Activation. *Journal of the Air & Waste Management Association*, Vol. 58, No. 1, (1958), pp. (78–84). ISSN 1047-3289.

[29] López, A & Márquez, C. (2003). Espectroscopía Infrarroja. In: *Técnicas de análisis y caracterización de materiales.* Faraldos, M., & Goberna, C. (Eds.). pp. (181-1859), CSIC, ISBN 84-00-08093-9.
[30] Lopez-Ramon, M.V., Stoeckli, F., Moreno-Castilla, C. & Carrasco-Marin, F. (1999). On the characterization of acidic and basic surface sites on carbons by various techniques. *Carbon*, Vol. 37, No. 8, (January 1999), pp. (1215-1221), ISSN 0008-6223.
[31] Maire, J. & Mering, J. (1970). Graphitization of soft carbons. In : *Chemistry and Physics of Carbon*. Walker, P.L. (ed.). pp. (125-189), Marcel Dekker, ISSN 0069-3138.
[32] Moreno-Castilla, C. (2004). Adsorption of organic molecules from aqueous solutions on carbon materials. *Carbon*, Vol. 42, No. 1, (2004), pp. (83-94), ISSN 0008-6223.
[33] Moreno-Castilla, C., López-Ramón, M.V. & Carrasco-Marín, F. (2000). Changes in surface chemistry of activated carbons by wet oxidation. *Carbon*, Vol. 38, No. 14, (2000), pp. (1995-2001), ISSN 0008-6223.
[34] Olivier, J. (1995). Modeling physical adsorption on porous and nonporous solids using density functional theory. *Journal of Porous Materials*, Vol. 2, No. 1, (June 1995), pp. (9-17), ISSN 1380-2224.
[35] Pastor-Villegas, J. & Durán-Valle, C.J. (2002). Pore structure of activated carbons prepared by carbon dioxide and steam activation at different temperatures from extracted rockrose. *Carbon*, Vol. 40, No. 3, (March 2002), pp. (397–402), ISSN 0008-6223.
[36] Pastor-Villegas, J., Durán-Valle, C.J., Valenzuela-Calahorro, C. & Gómez-Serrano, V. (1998). Organic chemical structure and structural shrinkage of chars prepared from rockrose. *Carbon*, Vol. 36, No. 9, (September 1998), pp. (1251–1256), ISSN 0008-6223.
[37] Pastor-Villegas, J., Gómez-Serrano, V., Durán-Valle, C.J. & Higes-Rolando, F.J. (1999). Chemical study of extracted rockrose and of chars and activated carbons prepared at different temperatures. *Journal of Analytical and Applied Pyrolysis*, Vol. 50, No. 1, (April 1999), pp. (1–16), ISSN 0165-2370.
[38] Pradhan, B.K. & Sandle, N.K. (1999). Effect of different oxidizing agent treatments on the surface properties of activated carbons. *Carbon*, Vol. 37, No. 8, (January 1999), pp. (1323-1332), ISSN 0008-6223.
[39] Raymundo-Piñero, E., Cazorla-Amorós, D., Linares-Solano, A., Find, J., Wild, U. & Schlögl, R. Structural characterization of N-containing activated carbon fibers prepared from a low softening point petroleum pitch and a melamine resin. *Carbon*, Vol. 40, No. 4, (April 2002), pp. (597-608), ISSN 0008-6223.
[40] Rubio-Montero, M.P., Durán-Valle, C.J., Jurado-Vargas, M., & Botet-Jiménez, A. (2009). Radioactive content of charcoal. *Applied Radiation and Isotopes*, Vol. 67, No. 5, (May 2009), pp. (953–956), ISSN 0969-8043.
[41] Scheele, C.W. (1780). *Chemical Observations and Experiments on Air and Fire*. J. Johnson, ISBN 9781171413929, London.
[42] Shen, W., Li, Z., & Liu, Y. (2008). Surface Chemical Functional Groups Modification of Porous Carbon. *Recent Patents on Chemical Engineering*, Vol. 1, No. 1, (January 2008), pp. (27-40), ISSN 2211-3347.
[43] Sing, K.S.W., Everett, D.H., Haul, R.A.W., et al. (1985). Reporting physisorption data for gas/solid systems with special reference to the determination of surface area and porosity (Recommendations 1984). *Pure and Applied Chemistry*, Vol. 57, No. 4, (1985), pp. (603-619), ISSN 1365-3075.

[44] Smith, B.C. (1999). *Infrared spectral interpretation: a systematic approach.* CRC Press, ISBN 0-8493-2463-7, Florida.

[45] Solum, M.S., Pugmire, R.J., Jagtoyen, M. & Derbyshire, F. (1995). Evolution of carbon structure in chemically activated wood. *Carbon*, Vol. 33, No. 9, (1995), pp. 1247-1254, ISSN 0008-6223.

[46] Somorjai, G.A. (1994). *Introduction to Surface Chemistry and Catalysis*, Wiley Interscience, ISBN 0471031925, UK.

[47] Stoeckli, H.F, Kraehenbuehl, F., & Morel, D. (1983). The adsorption of water by active carbons, in relation to the enthalpy of immersion. *Carbon*, Vol. 21, No. 6, (1983), pp. (589-591), ISSN 0008-6223.

[48] Valente-Nabais, J.M., & Carrott, P.J.M. (2006). Chemical Characterization of Activated Carbon Fibers and Activated Carbons. *Journal of Chemical Education*, Vol. 83, No. 3, (March 2006), pp. (436-438), ISSN 0021-9584.

[49] Washburn, E.W. (1921). The Dynamics of Capillary Flow. *Physical Reviews*, vol. 17, (1921), pp. (273-283).

[50] Yates, M. (2003). Área superficial, textura y distribución porosa. In: *Técnicas de análisis y caracterización de materiales.* Faraldos, M., & Goberna, C. (Eds.). pp. (221-246), CSIC, ISBN 84-00-08093-9.

[51] Zou, L., Huang, B., Huang, Y., Huang, Q., & Wang, Chang'an (2003). An investigation of heterogeneity of the degree of graphitization in carbon–carbon composites. *Materials Chemistry and Physics*, vol. 82, No.3, (December 2003), pp. (654–662), ISSN 0254-0584.

4

Applications of Activated Carbons Obtained from Lignocellulosic Materials for the Wastewater Treatment

Ma. del Rosario Moreno-Virgen, Rigoberto Tovar-Gómez, Didilia I. Mendoza-Castillo and Adrián Bonilla-Petriciolet
Instituto Tecnológico de Aguascalientes
México

1. Introduction

Activated carbons are used in a number of industrial applications including separation and purification technologies, catalytic processes, biomedical engineering, and energy storage, among others. The extensive application of activated carbon is mainly due to its relatively low-cost with respect to other adsorbents, wide availability, high performance in adsorption processes, surface reactivity and the versatility to modify its physical and chemical properties for synthesizing adsorbents with very specific characteristics (Haro et al., 2011). In particular, the adsorption on activated carbon is the most used method for wastewater treatment because it is considered a low-cost purification process where trace amounts of several pollutants can be effectively removed from aqueous solution. Recently, the demand of activated carbons has increased significantly as a water-purifying agent to reduce the environmental risks caused by the water pollution worldwide (Altenor et al., 2009; Bello-Huitle et al., 2010).

Traditionally, the activated carbons used in wastewater treatment are obtained from coal/lignite, wood or animal bones but, recently, there is a growing interest in the use of alternative and low-cost precursors for their production (Altenor et al., 2009; Elizalde-González & Hernández-Montoya, 2007; Mohamed et al., 2010). Specifically, lignocellulosic wastes are a low-cost natural carbon source for the synthesis of several materials including the production of activated carbons. In this context, it is convenient to remark that natural resources play a dominant role in the economic activities and the utilization of lignocellulosic wastes for the synthesis of valuable commercial products may contribute to the economic development and to prevent environment pollution especially in developing countries (Satyanarayana et al., 2007). Therefore, lignocellulosic materials are considered as an interesting and important natural resource for production of activated carbons based on the fact that several billion tons of these materials are available (Mohamed et al., 2010; Satyanarayana et al., 2007). Actually, these precursors are considered as the most appropriate candidates for a cost-effective preparation of activated carbons (Silvestre-Albero et al., 2012).

The production of activated carbons from lignocellulosic wastes is usually justified by two factors: the unique properties of these precursors and the possibility of mass production at an affordable cost. Several studies reported in literature indicate that it is possible to produce high quality activated carbons from raw lignocellulosic materials. These carbons are suitable for different applications including the removal of various pollutants from both drinking water and wastewaters. These pollutants include dyes, heavy metals, fluorides, phenols and other organic and inorganic toxic compounds, which are considered as priority substances for wastewater treatment in several countries (Altenor et al., 2009).

As indicated in Chapter 1, several raw lignocellulosic materials are available worldwide and they have been used in the synthesis of activated carbons for water purification. The performance of these activated carbons in the removal of a specific pollutant depends on both the surface chemistry and the textural properties (Altenor et al., 2009; Elizalde-González & Hernández-Montoya, 2007). Specifically, the properties of activated carbons such as the surface area, the pore size distribution, the presence of functional groups, and other physical and chemical parameters play an important role in the adsorption process of a given pollutant (Altenor et al., 2009). Previous chapters showed that these properties of activated carbon are mainly function of the precursor and of the type of thermal and activation process (Elizalde-González & Hernández-Montoya, 2007; Mohamed et al., 2010). As expected, activated carbons obtained from different precursors and synthesis procedures will show different adsorption properties for the removal of a given pollutant.

Considering the importance of these materials in the context of environmental pollution control, this chapter provides an overview of the application of activated carbons obtained from lignocellulosic precursors for wastewater treatment. This chapter summarizes studies reported mainly from the year 2000. Our analysis and discussion are focused on the performance of different activated carbons obtained from several precursors and their capabilities for the removal of relevant and toxic pollutants from water. The remainder of this chapter is organized as follows. In Section 2, we briefly introduce the activated carbon types obtained from different precursors and synthesis procedures used in the context of wastewater treatment. This discussion also comprises the use of several activated carbons for the removal of priority water pollutants. Section 3 provides the description of removal mechanisms, while some key aspects of carbon regeneration are reported in Section 4. Finally, in Section 5 we provide some remarks and conclusions about the application of lignocellulosic precursors for producing activated carbons for wastewater treatment and desalination.

2. Description of activated carbons for wastewater treatment and desalination

2.1 Synthesis, precursors and properties

To date, different lignocellulosic precursors have been used for the synthesis of activated carbons for water treatment. As stated in Chapter 2, activated carbons can be obtained from a variety of raw lignocellulosic materials and by different processing methods. Table 1 shows a summary of the most common precursors and synthesis conditions, including physical and chemical activation agents, used for the preparation of activated carbons for the removal of priority water pollutants. In general, a wide variety of activated carbons obtained from lignocellulosic precursor have been mainly used for the removal of heavy metals, dyes and phenol compounds.

		Common activating agent	
Pollutant	*Precursors*	*Physical*	*Chemical*
Heavy metals	*Arundo donax* plant canes, Apricot/Peach stones, Nut shells, *Havea braziliansis* sawdust, *Pinus sylvestris* sawdust, Peanut husk, Date pits, Oak cups pulp, *Terminalia arjuna* nut, *Casurina equisetifolia* leaves, *Spartina alterniflora* plant, Olive stone, Olive waste cakes, Cane sugar bagasse, African palm pit, Pecan nut shells, Chestnut shell, Grapeseed, Hazelnut shell, Coconut shells, Coir pith, Maize stalks	Air, CO_2, SO_2, H_2O, O_2	H_3PO_4, $AlCl_3$, HCl, H_2SO_4, KOH, NaOH, $ZnCl_2$, HNO_3, $FeCl_3$, $CaCl_2$, K_2CO_3
Dyes	Pecan nutshells, Castile nutshells, Mango seed, Guava seed, Mango husk, Orange seed, *Loofa egyptiaca* plant, Cane bagasse, Olive seeds, Rice straw, Bamboo, Coffee husk, Peanut hull, Pistachio shell, Oak cups pulp, Palm kernel shell, Apricot shell, Bagasse, Fir wood, Avocado kernel seed, *Posidonia oceanic (L)* dead leaves, Coconut shells	CO_2, H_2O, Air	H_3PO_4, HNO_3, $ZnCl_2$, KOH, $FeCl_3$, $CaCl_2$, K_2CO_3
Organic and inorganic pollutants (e.g., citric acid, phenol, para-nitrophenol; pharmaceuticals, fluorides)	*Havea Brazilians* sawdust, Oak cups pulp, Cedar wood, Fir wood, Brewer's spent grain lignin, Cashew nut shell, Pecan nuts, Peach stones, Kenaf, Rapessed, Almond Shell, Vine shoot, Oil palm fruit bunch	H_2O, CO_2	K_2CO_3, H_3PO_4, $ZnCl_2$, H_2SO_4, KOH, $ZrOCl_2$, HNO_3

Table 1. Lignocellulosic precursors and activating agents used in the synthesis of activated carbons for wastewater treatment and desalination.

It appears that the most studied precursors are bamboo, cane bagasse, olive wastes, plants and fruit wastes. It is important to remark that different activating agents can be used for improving the adsorption properties of these carbons and they include, for example, KOH, $ZnCl_2$, H_2SO_4, H_3PO_4, HNO_3 and CO_2. However, the chemical activation with both $ZnCl_2$ and H_3PO_4 appears to be the most used procedure for the preparation of adsorbents for water purification. As stated in Chapter 3, H_3PO_4 is a widely used activating agent for the preparation of activated carbons for wastewater treatment probably due to it can be removed easily after carbon activation by washing with hot and cold water besides its other operational and environmental advantages. On the other hand, the carbonization temperature usually used for the synthesis of these adsorbent ranges from 400 to 1000 °C. A summary of representative textural properties of some activated carbons are reported in Table 2. As expected, these activated carbons show a wide range of textural properties and surface chemistry which are determined by the conditions of the production process. For example, the studied activated carbons for wastewater treatment have surface areas from 100 to 1000 m^2/g.

Precursor	*Specific surface area, m²/g*	*Total pore volume, cm³/g*	*Mean pore diameter, nm*	*Reference*
Arundo donax plant canes	1043 - 1194	0.83 - 1.08	1.5 - 1.9	Basso et al. (2002)
Apricot stones	1249 - 1685	0.54 - 0.82	1.0 - 3.0	Puziy et al. (2007)
Bamboo	758 - 2471	0.42 - 1.34	2.2 - 3.4	Chan et al. (2009)
Cedar wood	846 - 1492	0.49 - 0.87	1.8 - 2.2	Cuerda-Correa et al. (2006)
Coir pith	167 - 910	0.12 - 0.36	1.6 - 2.8	Namasivayam & Sangeetha (2006)
Fit wood	891 - 2794	0.61 - 1.54	2.2 - 3.1	Wu et al. (2005)
Guava seed	52 - 598	0.12 - 0.50	3.3 - 8.9	Elizalde-González & Hernández-Montoya (2009)
Maize talks	196 - 1523	0.13 - 0.76	1.6 - 2.7	El-Hendawy (2009)
Mango husk	254 - 535	0.23 - 0.38	2.8 - 4.1	Elizalde-González & Hernández-Montoya (2008)
Orange seed	1 - 22	0.04 - 0.05	6.8 - 50.0	Elizalde-González & Hernández-Montoya (2008)
Pecan shell	27 - 1071	0.02 - 0.57	2.0	Dastgheib & Rockstraw (2001)

Table 2. Textural parameters of selected activated carbons obtained from lignocellulosic precursors for water treatment.

2.2 Applications in wastewater treatment and desalination

As indicated, most of activated carbons obtained from lignocellulosic precursors can be used for the removal of both organic and inorganic compounds especially heavy metals, dyes and phenol compounds. Until now, various adsorption kinetic and equilibrium experiments have been performed in the literature to study the adsorption process of priority water pollutants using lignocellulosic-based activated carbons. In general, these studies have been performed at batch conditions and a limited number of studies using dynamic conditions (i.e., packed bed columns) have been reported. With illustrative purposes, the application of several activated carbons for the adsorption of relevant water pollutants is discussed below.

2.2.1 Heavy metals

Water pollution by heavy metals is considered a serious environmental problem due to their toxicity, long persistence, bioaccumulation and biomagnifications in food chain (Wojnarovits et al., 2010). In fact, heavy metals are toxic to aquatic flora, animals and human beings even at relatively low concentrations (Tajar et al., 2009). Chromium, cadmium and lead are considered the most common hazardous metals found in wastewaters of several industries. Particularly, the adsorption of these heavy metals on activated carbons is the most used treatment method to reach water pollution regulations and environmental standards. A great variety of lignocellusic precursors and activating agents has been used for the preparation of activated carbons for heavy metal removal (see Table 3). In general, these precursors include plant canes, chestnut shell, cane sugar bagasse, coconut shell, maize talks, pecan nut, olive wastes and coir pith, among others. Both chemical and physical activation procedures have been used for the preparation of activated carbons for removal of metal ions. Some illustrative applications are described below.

Precursor	*Activating agent*	*Carbonization temperature, °C*	*Pollutant*	*Reference*
Arundo donax plant canes	H_3PO_4, Air, CO_2	500	Cadmium Nickel	Basso et al. (2002)
Nut shells	H_3PO_4, SO_2	475	Cadmium	Tajar et al. (2009)
Spartina alterniflora plant	H_3PO_4	400 - 700	Cadmium	Wang et al. (2011)
Apricot/Peach stones	H_3PO_4, Air	400 - 1000	Copper	Puziy et al. (2007)
Chestnut shell	$ZnCl_2$	550	Copper	Ozcimen & Ersoy-Mericboyu (2008)
Grapeseed	$ZnCl_2$	550	Copper	Ozcimen & Ersoy-Mericboyu (2008)
Pecan shell	Air, H_3PO_4	300 - 500	Copper	Dastgheib & Rockstraw (2001)
Olive wastes	H_3PO_4	350 - 650	Copper	Baccar et al. (2009)
Pinus sylvestris sawdust	HCl, H_3PO_4, NaOH, H_2SO_4	450 - 650	Chromium	Álvarez et al. (2007)
Oak cups pulp	H_3PO_4, $ZnCl_2$	600	Chromium	Timur et al. (2010)
Coconut palm	H_2O, O_2	800	Copper Cadmium Chromium	de Lima et al. (2011)
Cane sugar bagasse	HNO_3, Steam	900	Lead	Giraldo & Moreno-Piraján (2008)
African palm pit	HNO_3, Steam	900	Lead	Giraldo & Moreno-Piraján (2008)
Pecan nut shells	H_3PO_4, Calcium solution	800	Lead	Hernández-Montoya et al. (2011)
Maize talks	KOH	700	Lead	El-Hendawy (2009)
Coconut shell	$FeCl_3$, $ZnCl_2$, $CaCl_2$, K_2CO_3	500 - 800	Lead Copper Mercury	Gimba et al. (2009)
Coir pith	$ZnCl_2$	700	Nickel Mercury Chromium	Namasivayam & Sangeetha (2006)

Table 3. Synthesis conditions of activated carbons obtained from lignocellulosic precursors for heavy metal removal.

Chromium is considered a hazardous pollutant worldwide because of it is a mutagen and potential carcinogen. This pollutant is generated by several industries including metallurgy, leather tanning, and electroplating. In particular, water pollution caused by chromium is an important environmental problem in México and other developing countries. Until now, different lignocellulosic precursors have been used for the preparation of activated carbon for chromium removal. For example, Álvarez et al. (2007) reported the chromium adsorption using activated carbons obtained from *Pinus sylvestris* sawdust and different activation procedures. The activated carbons obtained from this precursor and different activating agents (i.e., HCl, H_3PO_4, H_2SO_4, $AlCl_3$, NaOH) showed different textural properties. This study suggested that the chromium adsorption may be controlled by a chemical or physical mechanism depending upon the used carbon activating agent. In general, activated carbons obtained with acid treatments exhibited the highest chromium adsorption especially those obtained by H_3PO_4 treatment. In other study, Timur et al. (2010) prepared activated carbons from oak cups pulp using H_3PO_4 and $ZnCl_2$ for chromium removal. These carbons have similar adsorption capacities, which are higher than 140 mg/g. These adsorption capacities are higher than those obtained from other lignocellusosic precursors treated with $ZnCl_2$ such as *Terminalia arjuna* nut (28 mg/g) and *Casurina equisetifolia* leaves (35 mg/g). Activated carbons prepared from coconut shell have been also used for chromium removal from water (de Lima et al., 2011). In particular, these adsorbents may show adsorption capacities up to 10 mg/g.

On the other hand, cadmium is also an important heavy metal in the context of environmental pollution control because its presence in water, even at very low concentrations, is harmful to aquatic environment and human health (Wang et al., 2011). This metal is also considered as a priority water pollutant in several countries. Several raw lignocellulosic materials have been employed as precursors for the preparation of activated carbons for cadmium adsorption. For example, Basso et al. (2002) studied the preparation of activated carbons from *Arundo donax* plant canes and phosphoric acid activation for the removal of cadmium and nickel ions. *Arundo donax L.* belongs to grass species and is a plant that attains heights of 40 ft and tends to form large, continuous root masses (Basso et al., 2002). This plant is considered as one of the most promising grass species for non-food uses because of it high biomass yield potential. Several activated carbons were prepared using this precursor at different atmospheres (i.e., air, CO_2 and N_2). This study concluded that activated carbons with the highest total content of surface oxygen functional groups showed the best adsorption capacity for heavy metal removal. A nut shell-based activated carbon obtained by chemical activation using H_3PO_4 and SO_2 has been also studied (Tajar et al., 2009). The surface modification of activated carbon from nutshells using these activation agents improves the cadmium adsorption properties. In other study, activated carbons obtained from coconut shells and different activating salts ($FeCl_3$, $ZnCl_2$, $CaCl_2$ and K_2CO_3) were employed for adsorption of cadmium ions (Gimba et al., 2009). In particular, adsorbents treated with $CaCl_2$ and K_2CO_3 showed the best removal performance. Alternatively, other precursors such as *Spartina alterniflora* plant (Wang et al., 2011) and coconut shell (de Lima et al., 2011) can be used for preparing adsorbents for cadmium removal.

Lead has been classified as a serious hazardous heavy metal because of it is extremely toxic for human beings. This metal is commonly detected in several industrial wastewaters from mining, smelting, metal plating and dying processes. Therefore, special attention has been given to develop proper adsorbents for lead removal from water. Specifically, Giraldo & Moreno-Piraján (2008) reported the adsorption of lead ions using activated carbons with high surface area obtained from cane sugar bagasse and African palm pit. These precursors were activated using HNO_3. In other study, activated carbons obtained from single step steam pyrolysis of sawdust of rubber wood were used for the simultaneous removal of lead and citric acid (Sreejalekshmit et al., 2009). Note that metal processing industries may discharge several pollutants including both heavy metals and organic chelating ligands such as citric acid and tartaric acid. Based on this fact, it is convenient to study the adsorption performance of activated carbons in multicomponent systems. These authors concluded that the presence of citric acid improves the adsorption capacity of this activated carbon. It appears that the functional -COOH groups of the adsorbed citric acid acted as new adsorption sites for lead on the activated carbon surface. Other adsorbents obtained from chemical activation of coconut shells (Gimba et al., 2009) and maize talks (El-Hendawy, 2009) can be used for lead adsorption. In particular, the synthesis of high performance activated carbons is feasible using maize talks and KOH activation. This adsorbent may show lead adsorption capacities up to 347 mg/g. Note that recent studies have showed that it is possible to use natural wastes (e.g., egg shell residues) to produce alternative and low-cost activating agents for improving the adsorption properties of activated carbons for heavy metal removal (Hernández-Montoya et al., 2011). One example is the activated carbon obtained from pecan nutshells, H_3PO_4 and a calcium solution extracted from egg shells, which have been used for the removal of lead ions.

Finally, removal studies for other heavy metals include the adsorption of mercury using coconut shell activated carbons (Gimba et al., 2009), the removal of nickel and mercury using $ZnCl_2$ activated coir pith carbon (Namasivayam & Sangeetha, 2006), and the copper adsorption employing carbons from pecan shell (Dastgheib & Rockstraw, 2001), fruit stones (Puziy et al., 2007), hazelnut shell (Dermibas et al., 2009), chestnut shell and grapeseed (Ozcimen & Ersoy-Mericboyu, 2009), olive-waste (Baccar et al., 2009) and coconut palm (de Lima et al., 2011).

		Conditions			
Precursor	*Pollutant*	*pH*	*T, °C*	q_{max}, *mg/g*	*Reference*
Arundo donax plant	Cadmium	5.8	28	57.3	Basso et al. (2002)
Nut shells	Cadmium	6	25	90 - 120	Tajar et al. (2009)
Pecan shell	Copper	2 - 5	25	33 - 40	Dastgheib & Rockstraw (2001)
Pinus sylvestris sawdust	Chromium	1 - 9	-	0.5 - 1.83	Álvarez et al. (2007)
African palm pit	Lead	2 - 8	25	4.7 - 15.2	Giraldo & Moreno-Piraján (2008)
Cane sugar bagasse	Lead	2 - 8	25	6.4 - 13.7	Giraldo & Moreno-Piraján (2008)
Maize talks	Lead	-	25	88 - 347	El-Hendawy (2009)
Pecan nut	Lead	5	30	75.4	Hernández-Montoya et al. (2011)
Arundo donax plant	Nickel	5.8	28	25.8	Basso et al. (2002)

Table 4. Adsorption capacities of selected activated carbons obtained from lignocellulosic precursors for heavy metal removal from water.

In general, literature indicates that the chemical activation using HNO_3 and H_3PO_4 improves the adsorption properties of activated carbons for heavy metal removal (Giraldo & Moreno-Piraján, 2008; Puziy et al., 2007). Specially, the presence of heteroatoms, in particular oxygen, may enhance the adsorption properties of these adsorbents. In summary, the adsorption capacities of lignocellulosic activated carbons for heavy metals may range from 10 to 100 mg/g depending on the precursor type and activation procedure. Adsorption capacities for selected lignocellulosic activated carbons are reported in Table 4.

2.2.2 Dyes

Water pollution by dyes is an important environmental problem because these pollutants are toxic and may be carcinogenic. Dyes can affect the physical and chemical properties of water and the aquatic flora and fauna. It has been estimated that about 7 X 10^5 tons of dyes are generated annually worldwide by several industrial activities. In general, dyes exhibit a wide range of different chemical structures and properties and can be classified according to chemical constitution, application and end use. Typical dyes used in industrial processes include acid, basic, direct, disperse and reactive (Dermibas, 2009). In general, dyes are usually resistant to classical biodegradation and the adsorption process is an effective treatment method for their removal (Demirbas et al., 2009). Table 5 reports some examples of activated carbons used for dye removal from water.

Precursor	*Activating agent*	*Carbonization temperature, °C*	*Pollutant*	*Reference*
Bamboo	H_3PO_4	400 - 600	Acid blue 25 Acid yellow 117	Chan et al. (2008) Chan et al. (2009)
Castile nutshells	H_3PO_4	500	Methylene blue	Bello-Huitle et al. (2010)
Coconut shell	$FeCl_3$, $ZnCl_2$, $CaCl_2$, K_2CO_3	500 - 800	Indigo blue	Gimba et al. (2009)
Fit wood	H_2O, KOH	780, 900	Acid blue 74 Basic brown 1 Methylene blue	Wu et al. (2005)
Loofa egyptiaca plant	H_3PO_4, HNO_3, $ZnCl_2$	500	Direct blue 106	El-Ashtoukhy (2009)
Oak cups pulp	H_3PO_4, $ZnCl_2$	600	Basic Red 18 Acid Red 111 Methylene Blue	Timur et al. (2010)
Pecan nutshells	H_3PO_4	500	Methylene blue	Bello-Huitle et al. (2010)
Pecan nut shells	H_3PO_4, Calcium solution	800	Acid blue 25	Hernández-Montoya et al. (2011)
P. oceanica (*L*) dead leaves	$ZnCl_2$	600	Methylene blue	Dural et al. (2011)

Table 5. Synthesis conditions of activated carbons obtained from lignocellulosic precursors for dye removal.

Recently, Altenor et al. (2009) have reported a review of activated carbons obtained from lignocellulosic wastes for water treatment giving an emphasis to those adsorbents used for dye removal. Literature review indicates that several adsorption studies using activated carbons for acid dyes have been reported. It is convenient to observe that antraquinone acid dyes (e.g., acid blue 25) are the second most important dyes commercially used in the textile industry and they are considered priority pollutants due to its high toxicity and carcinogenic potential (Hernández-Montoya et al., 2011). However, few adsorption studies have been performed using these acid dyes.

Woods are the most common used precursors for preparing activated carbons for dye removal. During long time, wood activated carbons obtained from different chemical and physical treatments have been used for the removal of several dyes including Acid Blue 74, Basic Brown 1, and Methylene Blue (Wu et al., 2005). Additionally, bamboo wastes have been widely employed as a raw material for the synthesis of activated carbons for dye removal (Ahmad & Hameed, 2010; Chan et al., 2008; Chan et al., 2009). Results of adsorption studies suggested that both surface area and porosity of these activated carbons play an important role in the removal process. Other precursors reported for dye removal include sugar industry wastes (Blanco-Castro et al., 2000), coir pith (Namasivayam & Sangeetha, 2006), and coconut shells (Gimba et al., 2009), among other raw lignocellulosic materials.

It is convenient to remark that several fruits and crops have been considered as effective precursors for preparation of adsorbents for dye removal. Specifically, Elizalde-González et al. (2007) reported the adsorption of several basic, acid and reactive dyes from aqueous solutions by avocado activated carbons impregnated with H_3PO_4. In other study, Elizalde-Gonzalez & Hernandez-Montoya (2008) reported the removal of anthraquinone dyes (i.e., acid blue 80 and acid green 27) using activated carbons obtained from seeds of mango, guava and orange. These fruits are abundant and are considered as a low cost agricultural waste in México and other countries (Elizalde-González & Hernández-Montoya, 2009). These studies indicated that the acid activated carbon obtained from orange seed was the most effective for the removal of acid green 27 from water despite its negligible specific surface; while the activated carbons obtained from mango seed were more effective for the removal of acid blue 80. Authors concluded that the interaction between the functional groups of activated carbons and these dyes are very important in the adsorption process. In other study, guava seed was used as precursor of activated carbon for the removal of Acid orange 7 (AO7), which is a dye commonly used in tanneries, paper manufacturing and textile industry (Elizalde-González & Hernández-Montoya, 2009). The optimal conditions for the preparation of guava seed carbon for dye removal were identified using a Taguchi experimental design. Recently, Bello-Huitle et al. (2010) reported the removal of methylene blue dye from water using activated carbons obtained from pecan and castile nutshells. Both pecan and castile nutshells are important crops from México and other countries. In this study, H_3PO_4 was used as activating agent for these precursors. Adsorption studies indicated that the adsorption capacity of activated carbon obtained from pecan nutshells was higher than that obtained for castile nutshells. Finally, Hernández-Montoya et al. (2011) have reported the preparation of activated carbons for adsorption of acid blue 25 using both pecan nut shells and a calcium solution extracted from egg shell wastes. This adsorbent showed adsorption capacities up to 48 mg/g.

With respect to precursors obtained from plant wastes, El-Ashtoukhy (2009) used the *Loofa egyptiaca* plant to prepare a low-cost activated carbon for the removal of direct blue 106. This plant is cultivated only in Egypt and is a highly branched, fibrous and interconnected cellulosic material. In this study, the adsorbents were obtained from the carbonization of this plant at 500 °C using H_3PO_4, HNO_3 and $ZnCl_2$ as activating agents. Adsorption kinetics and equilibrium studies were performed at 25 °C and different conditions of pH. Results showed that the maximum adsorption was obtained at pH 2. On the other hand, the sea plant *P. Occeanica* (*L*) can be also used as carbon precursor for methylene blue adsorption (Dural et al., 2011). In particular, kinetic and equilibrium adsorption studies of this dye were performed using an activated carbon obtained from this plant activated with $ZnCl_2$. Carbon samples showed adsorption capacities up to 280 mg/g at tested conditions.

Usually, the dye adsorption appears to be higher using activated carbons obtained from $ZnCl_2$ and H_3PO_4 activation. Adsorption capacities of lignocellulosic-based activated carbons may range from 50 to 400 mg/g. For illustration, dye adsorption capacities for selected activated carbons are reported in Table 6.

		Conditions			
Precursor	*Pollutant*	*pH*	*T*, °C	q_{max}, *mg/g*	*Reference*
Loofa egyptiaca plant	Direct blue 106	2 - 9	25	63.3 - 73.5	El-Ashtoukhy (2009)
Castile nutshells	Methylene blue	7	20	169.5	Bello-Huitle et al. (2010)
Pecan nutshells	Methylene blue	7	20	400	Bello-Huitle et al. (2010)
Pecan nut	Acid blue 25	6	30	48	Hernández-Montoya et al. (2011)

Table 6. Adsorption capacities of selected activated carbons obtained from lignocellulosic precursors for dye removal from water.

2.2.3 Phenol compounds

Phenol and its derivatives are common pollutants present in a variety of effluents from plastic, gasoline, disinfectant, pesticides, pharmaceutical, and steel industries (Lin et al., 2009; Timur et al., 2010). In particular, phenol is an important toxic compound listed as a priority pollutant by the EPA and other environmental protection agencies because of its high toxicity and possible accumulation in the environment. This compound is considered toxic to human beings and, as a consequence, it must be removed before the discharge of wastewaters. Until now, different adsorbents have been reported for the removal of phenol and its derivative compounds, see Table 7.

Specifically, batch experiments for phenol adsorption were performed using microporous activated carbons obtained from both kenaf and rapeseed precursors (Nabais et al., 2009). In particular, the phenol adsorption capacities of both adsorbents were higher than 70 mg/g. In other study, Alam et al. (2009) studied the effect of different synthesis conditions of activated carbon obtained from oil palm empty fruit bunches on phenol removal. These parameters include temperature, activation time and CO_2 flow rate. In general, these carbons showed phenol adsorption capacities from 1.03 to 4.83 mg/g. Recently, Bello-Huitle et al. (2010) studied the removal of phenol using activated carbons obtained by chemical activation and pyrolysis of pecan and castile nutshells. These authors synthesized different samples of activated carbons using these crops and adsorption isotherms were performed at 20 °C and pH 7. This study indicated that the activated carbon from pecan nutshells showed better phenol uptakes. On the other hand, Timur et al. (2010) reported that activated carbons obtained from oak cups pulp using $ZnCl_2$ have higher phenol adsorption capacities than those obtained for activated carbons produced with H_3PO_4. They suggested that this best performance was related to the lower amount of acidic surface groups on activated carbon.

Precursor	*Activating agent*	*Carbonization temperature, °C*	*Pollutant*	*Reference*
Castile nutshells	H_3PO_4	500	Phenol	Bello-Huitle et al. (2010)
Coir pith	$ZnCl_2$	700	Phenol	Namasivayam & Sangeetha (2006)
Kenaf	CO_2, HNO_3	700	Phenol	Nabais et al. (2009)
Oak cups pulp	H_3PO_4, $ZnCl_2$	600	Phenol	Timur et al. (2010)
Pecan nutshells	H_3PO_4	500	Phenol	Bello-Huitle et al. (2010)
Rapessed	CO_2, HNO_3	700	Phenol	Nabais et al. (2009)
Vine shoot	CO_2, HNO_3	800	Phenol p-nitrophenol	Mourao et al. (2011)
Almond shell	CO_2, HNO_3	800	Phenol p-nitrophenol	Mourao et al. (2011)
Cedar wood	H_2SO_4, CO_2	600 – 800	p-nitrophenol	Cuerda-Correa et al. (2006)
Fit wood	H_2O, KOH	780, 900	2,4 dichlorophenol 4-chlorophenol p-cresol, phenol	Wu et al. (2005)

Table 7. Synthesis conditions of activated carbons obtained from lignocellulosic precursors for removal of phenol compounds.

		Conditions			
Precursor	*Pollutant*	*pH*	*T, °C*	q_{max}, *mg/g*	*Reference*
Almond shell	Phenol	3	25	76 – 139	Mourao et al. (2011)
Castile nutshells	Phenol	7	20	53.2	Bello-Huitle et al. (2010)
Kenaf	Phenol	7.5	25	45 – 80	Nabais et al. (2009)
Pecan nutshells	Phenol	7	20	158.7	Bello-Huitle et al. (2010)
Rapeseed	Phenol	7.5	25	45 – 80	Nabais et al. (2009)
Vine shoot	Phenol	3	25	73	Mourao et al. (2011)
Almond shell	p-nitrophenol	3	25	154 – 224	Mourao et al. (2011)
Vine shoot	p-nitrophenol	3	25	126 – 238	Mourao et al. (2011)

Table 8. Adsorption capacities of selected activated carbons obtained from lignocellulosic precursors for removal of phenol compounds from water.

Finally, it is important to note that the adsorption of other phenol-based organic compounds on different types of activated carbons has been also studied and they include, for example, the removal of p-nitrophenol using activated carbons obtained from cedar wood activated by H_2SO_4 (Cuerda-Correa et al., 2006), carbons from almond shell and vine shoot precursors activated with carbon dioxide and oxidized with nitric acid (Mourao et al., 2011), the adsorption of phenol-compounds using coir pith activated carbons (Namasivayam & Sangeetha, 2006), or the removal of 2,4-dichloropenol, 4-chloropenol and p-cresol using KOH and steam-activated carbons obtained from Fir wood (Wu et al., 2005). Overall, these activated carbons showed adsorption capacities from 40 to 200 mg/g. Table 8 shows the adsorption capacities of different activated carbons used for removal of phenol compounds.

2.2.4 Other organic and inorganic toxic pollutants

Activated carbons can be used for the removal of other organic and inorganic compounds such as nitrates, thyocyanate, selenite, vanadium, sulfates, molybdate, fluorides, and pharmaceuticals, among other pollutants (Namasivayam & Sangeetha, 2006).

Drinking water sources in a number of developing and underdeveloped countries is polluted by toxic anions such as fluorides. Specifically, the presence of high fluoride concentrations in drinking water is a common problem in several countries including México, China and India (Alagumuthu & Rajan, 2010; Hernández-Montoya et al., 2011). Fluoride concentrations in drinking water higher than 1.5 mg/L is the principal cause of dental fluorosis in children and may cause bone fluorosis if a chronic exposure occurs. Traditional lignocellulosic precursors and activation procedures are not suitable for the production of activated carbons suitable for fluoride removal from water. However, some studies have shown that fluoride ion has a strong affinity towards multivalent metal ions, e.g., Al^{3+}, Fe^{3+} and Zr^{4+} (Alagumuthu & Rajan, 2010) and also may interact with some bivalent ions such as calcium (Hernández-Montoya et al., 2011). Based on this fact, some studies have reported the application of non-conventional impregnating agents for the preparation of activated carbons for fluoride removal from water. For example, Alagumuthu & Rajan (2010) studied the carbonization of cashew nut shell impregnated with zirconium oxy chloride. Cashew nut is one of the commercialized products of the cashew tree and the cashew nut shell is the waste product of cashew nut, which contains potassium and magnesium compounds. In general, adsorption capacities of this carbon were around 2.0 mg/g at tested conditions. These authors indicated that the fluoride adsorption was related to both electrostatic interactions and a chemisorption mechanism that involves chloride and hydroxide species of this carbon. In particular, the presence of zirconium species improves the adsorption properties of this adsorbent. A recent study showed that activated carbons obtained from pecan nut shells and egg shell wastes can be used for fluoride removal from water (Hernández-Montoya et al., 2012). This study concluded that the calcium chemical species on the carbon surface played an important role in the fluoride adsorption process.

On the other hand, activated carbons have been used for the removal of pharmaceuticals from water (e.g., Cabrita et al., 2010). Several pharmaceuticals are released to the environment via human and animal excreta and, as a consequence, trace quantities of these pollutants tend to accumulate in water resources. Cabrita et al. (2010) reported the removal of acetaminophen (i.e., paracetamol) from aqueous solution using an activated carbon from peach stones. This activated carbon is characterized by a high amount of oxygen functionalities, which appear to be related to the presence of pyrone and/or chromene-like type structures. This carbon showed an adsorption capacity higher than those obtained for activated carbon synthesized from plastic waste and commercial carbons. This study concluded that the adsorption of this pharmaceutical is a complex process that depends on both the chemical composition and the textural parameters of activated carbon.

Recently, the performance of activated carbon has been studied and tested in the removal of new environmental pollutants originated from consumer products and by-products used in industrial, agricultural and other human activities. In particular, these emergent pollutants include pesticides, household-cleaning chemicals, fragrances, and disinfectants, among other organic and inorganic toxic compounds (Cabrita et al., 2010). Based on this perspective, it is expected that the applications of activated carbons obtained from lignocellulosic precursors will increase for wastewater treatment in forthcoming years.

3. Description of adsorption mechanisms of priority water pollutants using activated carbons

The surface chemistry of activated carbons plays an important role to determine their adsorption performance in wastewater treatment. In particular, the precursor has a critical effect on the surface chemistry properties of activated carbons (Wang et al., 2011). The presence of several functional groups on carbon surface (e.g., carboxylic, carbonyl, hydroxyl, ether, quinine, lactone, anhydride) implies the presence of many types of pollutant-carbon interactions (Wu et al., 2005). It is important to remark that the nature and prevalence of functional groups on carbon surface may be modified by activation methods.

Several studies have shown that the surface functional groups of activated carbons play an important role for the adsorption of a specific pollutant (Ould-Idriss et al., 2011). For example, polar or acidic oxygen functional groups on the surface of activated carbons have been recognized to play a fundamental role on metal adsorption (Basso et al., 2002). In fact, reported studies have shown the predominant influence of surface oxygen functional groups of activated carbons on metal uptake (Basso et al., 2002). For this application, the following trend has been identified: the higher the content of functional groups, the greater the adsorption extent of the activated carbon. These functional groups include: carbonyls, phenols, lactones and carboxylic acids (Basso et al., 2002). Also, activated carbons with sulfur functional groups are suitable for the removal of some heavy metals such as cadmium (Tajar et al., 2009). In the case of phenol and its derivatives compounds, the adsorption process is also related to the oxygen-containing surface functional groups (Tamir et al., 2010). Literature indicates that the most relevant heteroatoms for phenol adsorption are nitrogen and oxygen (Nabais et al., 2009).

On the other hand, cation-exchange mechanisms are also involved in the adsorption of some pollutants from water. Some studies have suggested that the cation-exchange properties of activated carbons are determined by the presence of oxygen- and phosphorous-containing surface groups. For example, Dastgheib & Rockstraw (2001) reported that an ion-exchange and surface complexation with oxygen- and phosphorus-containing groups on pecan shell activated carbon may be involved in the adsorption of copper from water. This cation exchange capacity of activated carbons can be improved via chemical activation (Puziy et al., 2007). Usually, activated carbons obtained from carbonization and phosphoric acid activation may show a considerable cation exchange capacity.

Electrostatic interactions appear to play a key role in the adsorption mechanism of some pollutants including metals ions and dyes (Wang et al., 2011). For the case of some dyes, the adsorption process has been related to electrostatic force of attraction between dyes and activated carbon and also to complex formation (Tamir et al., 2010). In general, literature indicates that dye removal using activated carbons may imply several mechanisms including ion-dipole forces, ion exchange, hydrogen bonding and non-specific interactions. Also, phenol adsorption onto activated carbon may occur via a complex interplay of electrostatic and dispersion interactions (Nabais et al., 2009).

It is convenient to recall that the adsorption process using activated carbon depends on the operating conditions of removal process (i.e., adsorbent mass, pollutant concentration, operating mode), the adsorbent characteristics (e.g., functional groups, textural properties, etc.) and the solution chemistry (e.g., temperature, pH, ionic strength). Based on this fact, diverse adsorption mechanisms may occur simultaneously during the removal of a specific pollutant using activated carbons synthesized from lignocellulosic precursors. Therefore, the identification and characterization of adsorption mechanisms involved in the removal of priority water pollutants is a relevant and important research topic for understanding the chemistry of activated carbons.

4. Desorption and regeneration studies of activated carbons used in water treatment

Adsorbent regeneration is an important operating parameter to establish the feasibility and the operating costs of water treatment processes using activated carbons. The performance and efficacy of desorbing agents depends on the carbon type, the concentration of sorbed pollutant, and the operating conditions of desorption process (e.g., concentration of desorbing agent, temperature and sorbent dosage).

In general, literature indicates that some regeneration studies have been performed using activated carbons obtained from lignocellulosic precursors loaded with some waster pollutants especially heavy metal ions. Specifically, copper desorption from pecan shell activated carbon has been studied using both water and 10% HCl solution (Dastgheib & Rockstraw, 2001). This study showed that copper desorption with water is not feasible, while HCl may recovery up to 98% of the copper adsorbed on pecan shell activated carbon. Basso et al. (2002) performed desorption studies to recovery cadmium and nickel ions adsorbed on activated carbons obtained from *Arundo donax* plant canes using HCl. This study concluded that it is feasible to recovery the 90% of the metal ions loaded in these activated carbons. In other study, Tajar et al. (2009) reported preliminary results for desorption of cadmium from an activated carbon obtained from nutshells using HCl, HNO_3 and H_2SO_4 as extractants. It appears that HCl is an effective chemical for desorbing cadmium ions from this activated carbon. Authors concluded that H^+ ions from HCl displace cadmium ions bounded to the activated carbon during the desorption stage. Recently, the chromium desorption from activated carbon obtained from *Pinus sylvestris* sawdust has been studied using H_2SO_4 (Álvarez et al., 2007). Results of carbon regeneration showed that this activated carbon retains its chromium adsorption capacity during the first regeneration cycle. However, the adsorption performance of this adsorbent is substantially reduced in subsequent regeneration cycles. With respect to other pollutants, HCl and NaOH have been used for the fluoride desorption using a zirconium impregnated cashew nut shell carbon (Alagumuthu & Rajan, 2010). These authors reported that NaOH was more effective for carbon regeneration and may recovery more than 95% of fluoride loaded on activated carbon.

In summary, a limited number of studies have been reported for the regeneration of activated carbons used for wastewater treatment. Therefore, further research should be performed to develop low-cost and effective regeneration procedures for activated carbons used in water purification.

5. Conclusions

This chapter describes the application of lignocellulosic precursors for the synthesis of activated carbons used in the removal of different pollutants from drinking water and wastewaters. In particular, lignocellulosic precursors can be used for the synthesis of activated carbons with attractive properties for the adsorption of different organic and inorganic pollutants. Literature indicates that we can prepare activated carbons with improved adsorption properties to remove effectively priority water pollutants by using the appropriate lignocellulosic precursors and by optimizing the conditions of carbonization and activation. In particular, research on activated carbon for wastewater treatment should give special attention in the optimization of synthesis conditions for improving adsorption properties to remove hazardous pollutants such as fluoride and arsenic. Also, the development of low-cost regeneration procedures is highlighted to reduce the costs of water treatment technologies. Finally, it is expected that the applications of activated carbons obtained from these precursors will increase for wastewater treatment and other science fields in forthcoming years.

6. Acknowledgments

Authors acknowledge the financial support provided by CONACYT, DGEST and Instituto Tecnológico de Aguascalientes (México).

7. References

[1] Ahmad, A.A. & Hameed, B.H. (2010). Fixed-bed adsorption of reactive azo dye onto granular activated carbon prepared from waste. *Journal of Hazardous Materials*, Vol. 175, No. 1-3 (March 2010), pp. (298–303), ISSN 0304-3894.

[2] Alagumuthu, G. & Rajan M. (2010). Equilibrium and kinetics of adsorption of fluoride onto zirconium impregnated cashew nut shell carbon. *Chemical Engineering Journal*, Vol. 158, No. 3 (April 2010), pp. (451-457), ISSN 1385-8947.

[3] Alam, M.Z., Ameem, E.S., Muyibi, S.A. & Kabbashi, N.A. (2009). The factors affecting the performance of activated carbon prepared from oil palm empty fruit bunches for adsorption of phenol. *Chemical Engineering Journal*, Vol. 155, No. 1-2, (February 2011), pp. (191-198), ISSN 1385-8947.

[4] Álvarez, P., Blanco, C. & Granda, M. (2007). The adsorption of chromium (VI) from industrial wastewater by acid and base-activated lignocellulosic residues. *Journal of Hazardous Materials*, Vol. 144, No. 1-2, (June 2007), pp. (400-405), ISSN 0304-3894.

[5] Altenor, S., Carene-Melane, B. & Gaspard, S. (2009). Activated carbons from lignocellulosic waste materials for water treatment: a review. *International Journal of Environmental Technology and Management*, Vol. 10, No. 3-4, pp. (308-326), ISSN 1466-2132.

[6] Baccar, R., Bouzid, J., Feki, M. & Montiel A. (2009). Preparation of activated carbon from Tunisian olive-waste cakes and its application for adsorption of heavy metal ions. *Journal of Hazardous Materials*, Vol. 162, No. 2-3, (March 2009), pp. (1522–1529), ISSN 0304-3894.

[7] Basso, M.C., Cerrella, E.G. & Cukierman A.K. (2002). Activated carbons developed from rapidly renewable bioresource for removal of cadmium (II) and nickel (II) ions from dilute aqueous solution. *Industrial Engineering Chemical Research*, Vol. 41, No. 2, (December 2001), pp. (180-189), ISSN 0888-5885.

[8] Bello-Huitle, V., Atenco-Fernández, P. & Reyes-Mazzoco, R. (2010). Adsorption studies of methylene blue and phenol onto pecan and castile nutshells prepared by chemical activation. *Revista Mexicana de Ingeniería Química*, Vol. 9, No. 3, pp. (313-322), ISSN 1665-2738.

[9] Blanco-Castro, J., Bonelli, P., Cerrella E. & Cukierman A.L. (2000). Phosphoric acid activation of agricultural residues and bagasse from sugar cane: influence of the experimental conditions on adsorption characteristics of activated carbons. *Industrial Engineering Chemical Research*, Vol. 39, No. 11, (September 2000), pp. (4166-4172), ISSN 0888-5885.

[10] Cabrita, I., Ruiz, B., Mestre, A.S., Fonseca, I.M., Carvalho, A.P. & Ania C.O. (2010). Removal of an analgesic using activated carbons prepared from urban and industrial residues. *Chemical Engineering Journal*, Vol. 163, No. 3, (October 2010), pp. (249-255), ISSN 1385-8947.

[11] Chan, L.S., Cheung, W.H. & McKay, G. (2008). Adsorption of acid dyes by bamboo derived activated carbon. *Desalination*, Vol. 218, No. 1-3, (January 2008), pp. (304–312), ISSN 0011-9164.

[12] Chan, L.S., Cheung, W.H. Allen, S.J. & McKay G. (2009). Separation of acid-dyes mixture by bamboo derived active carbon. *Separation and Purification Technology*, Vol. 67, No. 2, (June 2009), pp. 166-172, ISSN 1383-5866.

[13] Cuerda-Correa, E.M., Díaz-Díez, M.A., Macías-García, A. & Gañán-Gómez, J. (2006). Preparation of activated carbons previously treated with sulfuric acid. A study of their adsorption capacity in solution. *Applied Surface Science*, Vol. 252, No. 17, (June 2006), pp. (6042-6045), ISSN 0169-4332.

[14] Dastgheib, S.A. & Rockstraw, D.A. (2001). Pecan shell activated carbons: synthesis, characterization, and application for the removal of copper from aqueous solution. *Carbon*, Vol. 39, No. 12, (October 2001), pp. (1849-1855), ISSN 0008-6223.

[15] de Lima, L.S., Machado-Araujo, M.D., Quináia, S.P., Migliorine, D.W. & Garcia, J.R. (2011). Adsorption modeling of Cr, Cd and Cu on activated carbon of different origins by using fractional factorial design. *Chemical Engineering Journal*, Vol. 166, No. 3, (February 2011), pp. (881–889), ISSN 1385-8947.

[16] Demirbas, A. (2009). Agricultural based activated carbons for the removal of dyes from aqueous solutions: A review. *Journal of Hazardous Materials*, Vol. 167, No. 1-3, (August 2009), pp. (1-9), ISSN 0304-3894.

[17] Dermibas, E., Dizge, N., Sulak, M.T. & Kobya, M. (2009). Adsorption kinetics and equilibrium studies of copper from aqueous solutions using hazelnut shell activated carbon. *Chemical Engineering Journal*, Vol. 148, No. 2-3, (May 2009), pp. (480-487), ISSN 1385-8947.

[18] Dural, M.U., Cavas, L., Papageorgious, S.K. & Katsaros, F.K. (2011). Methylene blue adsorption on activated carbon prepared from *Posidonia oceánica (L)* dead leaves: kinetics and equilibrium studies. *Chemical Engineering Journal*, Vol. 168, No. 1, (March 2011), pp. (77-85), ISSN 1385-8947.

[19] El Ashtoukhy E.S.Z. (2009). Loofa egyptiaca as a novel adsorbent for removal of direct blue dye from aqueous solution. *Journal of Environmental Management*, Vol. 90, No. 8, (June 2009), pp. (2755–2761), ISSN 0301-4797.

[20] El-Hendawy, A.N.A. (2009). An insight into the KOH activation mechanism through the production of microporous activated carbon for the removal of Pb^{2+} cations. *Applied Surface Science*, Vol. 255, No. 6, (January 2009), pp. (3723–3730), ISSN: 0169-4332.

[21] Elizalde-González, M.P. Mattusch, J. Peláez-Cid, A.A. & Wennrich, R. (2007). Characterization of adsorbent materials prepared from avocado kernel seeds: natural, activated and carbonized forms. *Journal of Analytical and Applied Pyrolysis*, Vol. 78, No. 1, (January 2007), pp. (185-193), ISSN 0165-2370.

[22] Elizalde-González, M.P. & Hernández-Montoya, V. (2008). Fruit seeds as adsorbents and precursors of carbon for the removal of anthraquinone dyes. *International Journal of Chemical Engineering*, Vol. 1, No. 2-3, pp. (243-253), ISSN 0974-5793.

[23] Elizalde-González, M.P. & Hernández-Montoya, V. (2009). Removal of acid orange 7 by guava seed carbon: A four parameter optimization study. *Journal of Hazardous Materials*, Vol. 168, No. 1, (August 2009), pp. (515-522), ISSN 0304-3894.

[24] Elizalde-González, M.P. & Hernández-Montoya, V. (2009). Guava seed as an adsorbent and as a precursor of carbon for the adsorption of acid dyes. *Bioresource Technology*, Vol. 100, No. 7, (April 2009), pp. (2111-2117), ISSN 0960-8524.

[25] Gimba, C.E., Turoti, M., Egwaikhide, P.A. & Akporhonor, E.E. (2009). Adsorption of indigo blue dye and some toxic metals by activated carbons from coconut shells. *Electronic Jorunal of Environmental, Agricultural and Food Chemistry*, Vol. 8, No. 11, pp. (1194-1201), ISSN 1579-4377.

[26] Giraldo, L. & Moreno-Piraján, J.C. (2008). Pb^{2+} adsorption from aqueous solutions on activated carbons obtained from lignocellulosic residues. *Brazilian Journal of Chemical Engineering*, Vol. 25, No. 1, (January 2008), pp. (143-151), ISSN 0104-6632.

[27] Haro, M., Ruiz, B., Andrade, M., Mestre, A.S., Parra, J.B., Carvalho, A.P. & Ania, C.O. (2011). Dual role of copper on the reactivity of activated carbons from coal and lignocellulosic precursors. *Microporous and Mesoporous Materials*, In press, ISSN 1387-1811.

[28] Hernández-Montoya, V., Mendoza-Castillo, D.I., Bonilla-Petriciolet, A., Montes-Morán, M.A. & Pérez-Cruz, M.A. (2011). Role of the pericarp of Carya illinoinensis as biosorbent and as precursor of activated carbon for the removal of lead and acid blue 25 in aqueous solutions. *Journal of Analytical and Applied Pyrolysis*, Vol. 92, No. 1, (September 2011), pp. (143-151), ISSN 0165-2370.

[29] Hernández-Montoya, V., Ramírez-Montoya, L.A., Bonilla-Petriciolet, A. & Montes-Moran, M. (2012). Optimizing the removal of fluoride from wáter using new carbons obtained by modification of nut shell with a calcium solution from egg shell. *Biochemical Engineering Journal*, In press, ISSN 1369-703X.

[30] Krishnan, K.A., Sreejalekshmi, K.G. & Varghese, S. (2010). Adsorptive retention of citric acid onto activated carbon prepared from *Havea braziliansis* sawdust: Kinetic and isotherm overview. *Desalination*, Vol. 257, No. 1-3, (July 2010), pp. (46-52), ISSN 0011-9164.

[31] Mohamed, A. R., Mohammadi, M. & Darzi, G.N. (2010). Preparation of carbon molecular sieve from lignocellulosic biomass: A review. *Renewable and Sustainable Energy Reviews,* Vol. 14, No. 6, (August 2010), pp. (1591-1599), ISSN 1364-0321.

[32] Mourao, P.A.M., Laginhas, C., Custodio, F., Nabais, J.M.V., Carrot, P.J.M. & Ribeiro Carrott, M.M.L. (2011). Influence of oxidation process on the adsorption capacity of activated carbon from lignocellulosic precursors. *Fuel Processing Technology,* Vol. 92, No. 2, (February 2011), pp. (241-246), ISSN 0378-3820.

[33] Mussatto, S.I., Fernandez, M., Rocha, G.J.M., Orfao, J.J.M., Teixeira J.A. & Roberto, I.C. (2010). Production, characterization and application of activated carbon from brewer's spent grain lignin. *Bioresource Technology,* Vol. 101, No. 7, (April 2010), pp. (2450-2457), ISSN 0960-8524.

[34] Namasivayam, C. & Sangeetha, D. (2006). Recycling of agricultural solid waste, coir pith: Removal of anions, heavy metals, organics and dyes from water by adsorption onto $ZnCl_2$ activated coir pith carbon. *Journal of Hazardous Materials,* Vol. 135, No. 1-3 (July 2006), pp. (449–452), ISSN 0304-3894.

[35] Ould-Idriss, A., Stitou, M., Cuerda-Correa, E.M., Fernández-González, C., Macías-García, A., Alexandre-Franco, M.F. & Gómez-Serrano, V. (2011). Preparation of activated carbons from olive-tree Wood revisited. II. Physical activation with air. *Fuel Processing Technology,* Vol. 92, No. 2, (February 2011), pp. (266-270), ISSN 0378-3820.

[36] Ozcimen, D. & Ersoy-Mericboyu, A. (2009). Removal of copper from aqueous solutions by adsorption onto chestnut shell and grapeseed activated carbons. *Journal of Hazardous Materials,* Vol. 168, No. 2-3, (September 2009), pp. (1118-1125), ISSN 0304-3894.

[37] Puziy, A.M.; Poddubnaya, O.I.; Martínez-Alonso, A.; Castro-Muñiz, A.; Suárez-García, F. & Tascón, J.M.D. (2007). Oxygen and phosphorus enriched carbons from lignocellulosic material. *Carbon,* Vol. 45, No. 10, (September 2007), pp. (1941-1950), ISSN 0008-6223.

[38] Satyanarayan, K.G., Guimaraes, J.L. & Wypych, F. (2007). Studies on lignocellulosic fibers of Brazil. Part I: source, production, morphology, properties and applications. *Composites: Part A,* Vol. 38, No. 7, (July 2007), pp. (1694-1709), ISSN 1359-835X.

[39] Silvestre-Albero, A., Goncalvez, M., Itoh, T., Kaneko, K., Endo, M., Thommes, M., Rodríguez-Reinoso, F. & Silvestre-Albero, J. (2012). Well-defined mesoporosity on lignocellulosic-derived activated carbons. *Carbon,* Vol. 50, No. 1, (January 2012), pp. (66-72), ISSN 0008-6223.

[40] Sreejalekshmi, K.G., Anoop-Krishnan, K. & Anirudhan, T.S. (2009). Adsorption of Pb(II) and Pb(II)-citric acid on sawdust activated carbon: kinetic and equilibrium isotherm studies. *Journal of Hazardous Materials,* Vol. 161, No. 2-3, (January 2009), pp. (1506-1513), ISSN 0304-3894.

[41] Tajar, A.F., Kaghazchi, T. & Soleimani, M. (2009). Adsorption of cadmium from aqueous solutions on sulfurized activated carbon prepared from nut shells. *Journal of Hazardous Materials,* Vol. 165, No. 1-3, (June 2009), pp. (1159-1164), ISSN 0304-3894.

[42] Timur, S., Kantarli, I.C., Onenc, S. & Yanik, J. (2010). Characterization and application of activated carbon produced from oak cups pulp. *Journal of Analytical and Applied Pyrolysis,* Vol. 89, No. 1, (September 2010), pp. (129-136), ISSN 0165-2370.

[43] Valente Nabis, J.M., Gomez, J.A., Suhas, Carrott, P.J.M., Laginhas, C. & Roman, S. (2009). Phenol removal onto novel activated carbons made from lignocellulosic precursors: influence of surface properties. *Journal of Hazardous Materials,* Vol. 167, No. 1-3, (August 2009), pp. (904-910), ISSN 0304-3894.

[44] Wang, Z., Nie, E., Li J., Zhao Y., Luo X. & Zheng Z. (2011). Carbons prepared from Spartina alterniflora and its anaerobically digested residue by H_3PO_4 activation: Characterization and adsorption of cadmium from aqueous solutions. *Journal of Hazardous Materials*, Vol. 188, No. 1-3, (April 2011), pp. (29-36), ISSN 0304-3894.

[45] Wojanárovits, L., Földváry, Cs.M. & Takács E. (2010). Radiation-induced grafting of cellulose for adsorption of hazardous water pollutans: A review. *Radiation Physics and Chemistry*, Vol. 79, No. 8, (August 2010), pp. (848-862), ISSN 0969-806X.

[46] Wu, F.C., Tseng, R.L. & Juang, R.S. (2005). Preparation of highly microporous carbons from fir wood by KOH activation for adsorption of dyes and phenols from water. *Separation and Purification Technology*, Vol. 47, No. 1-2, (December 2005), pp. (10-19), ISSN 1383-5866.

Characterization of Pyrolysis Products Obtained During the Preparation of Bio-Oil and Activated Carbon

Rosa Miranda, César Sosa, Diana Bustos, Eileen Carrillo and María Rodríguez-Cantú
Universidad Autónoma de Nuevo León
México

1. Introduction

Nowadays, energy security and sustainable development are two major challenges encountered by the world. Renewable energy should be studied extensively to explore new technologies and in order to maintain secure energy sources for sustainable development, considering the fact that the energy demand is increasing, depleting fossil fuel reserves, with increasing populations and economic development.

Biomass is one of the most important renewable energy sources and is considered an alternative to fossil fuels. Biomass thermo chemical conversion processes including pyrolysis, combustion, gasification and liquefaction are employed for power generation and production of liquid biofuels, chemicals and charcoal, which can be used as activated carbon. Biomass is mainly composed of carbon; recently this property has been very attractive for the purpose of producing functional carbon materials, which have relevant economic and environmental implications.

Biomass resources include wood from plantation forests, residues from agricultural or forest production, and organic waste by-products from industry, domesticated animals, and human activities. The chemical energy contained in the biomass is derived from solar energy using the process of photosynthesis. This is the process by which plants take in carbon dioxide and water, using energy from sunlight, convert them into sugars, starches, cellulose, lignin etc., and finally oxygen is produced and released.

Pyrolysis of biomass is a promising method for simultaneous production of activated carbon, bio-oil and gaseous fuels and other valuable chemicals, while the almost simultaneous pyrolysis and gasification of the fuel result in formation of solid product with high surface area and well-developed porous structure (Nickolov & Mehandjiev, 1995; Mehandjiev et al., 1997). Pyrolysis is the thermal destruction of organic macromolecules in the absence of oxygen in small molecules. The destructed portion comprises a high energy content and significant organic content, which leads to the possibility of energy extraction as well as the production of activated carbon and chemicals from biomass (Prakash & Karunanithi, 2008).

As stated in previous chapters, activated carbons are carbonaceous materials with a high surface area and porous structure, sometimes described as solid sponges (Abdel-Nasser & El-Hendawy, 2005). The large surface area results in a high capacity for adsorbing chemicals from gases or liquids. Activated carbons are versatile adsorbents with a wide range of applications such as adsorbents for treatment and purification of water, air as well as various chemical and natural products (Abdel-Nasser & El-Hendawy, 2005; Budinova et al., 2006). The increasing use of activated carbon is due to the necessity of environmentally friendly processes and also for material recovery purposes.

Chapter 1 shows that different types of biomass materials and waste products have been studied for activated carbon production. These precursors include wood (Ahmad et al., 2006), coal (Lozano-Castello et al., 2005), nut shells (Lua et al., 2004), husks (Baquero et al.,2003), and agricultural by-products (Abdel-Nasser & El-Hendawy, 2005; Durán-Valle et al., 2005; Budinova et al., 2006). In addition to the use as an adsorbent, high porosity carbons have been recently applied in the manufacture of high-performance layer capacitors. Because of the introduction of rigorous environmental regulations and the development of new applications, the demand for porous carbons is expected to increase progressively (Sircar et al., 1996). Applications of pyrolysis products have some disadvantages due to the high degree of heterogeneity in their form and composition. Characteristics of these products depend on the operating conditions and the type of biomass used, so it requires more knowledge of the conversion process.

Pyrolysis is discussed here to improve the valorisation of two Mexican typical agricultural wastes for energy and carbon activated production. The product characteristics, their relative proportions in the gas/liquid/solid phases and the process energy requirements depend upon the input material and the process conditions. Therefore, the goal of this chapter is to describe the conversion of waste biomass into activated carbon. Waste biomass like orange peel and pecan nut shell is converted thermally in one step. First, the biomass undergoes a pyrolysis process at 600 °C in nitrogen atmosphere. The gaseous and liquid pyrolysis products were collected as bio-oil, and then they can be used as fuel either for heating the facilities or for electricity production.

2. Experimental method for biomass pyrolysis

Bench scale experiments were carried out in a pyrolysis system with controlled temperature and a semi-batch stainless steel reactor. The schematic diagram of the process is illustrated in Figure 1. The reactor has a volume of three litters, and is externally heated by an electrical furnace. Pyrolysis experiments are normally performed with approximately 400 g of feedstock. The sample was placed inside the reactor and heated at 600 °C for one hour. The gases and vapours generated during pyrolysis pass through a condensation train, which consists of four Pyrex traps. The remaining non-condensable gases are collected and stored in a plastic sampling bag with a valve for future chromatography analysis. Pyrolysis product yields are determined by weighing the char and bio-oil. Non-condensable gases yield is calculated by the mass difference. Pyrolysis end temperatures were fixed at 600, 700 and 750 °C. All experiments were performed under nitrogen atmosphere using a flow of 60 ml/min.

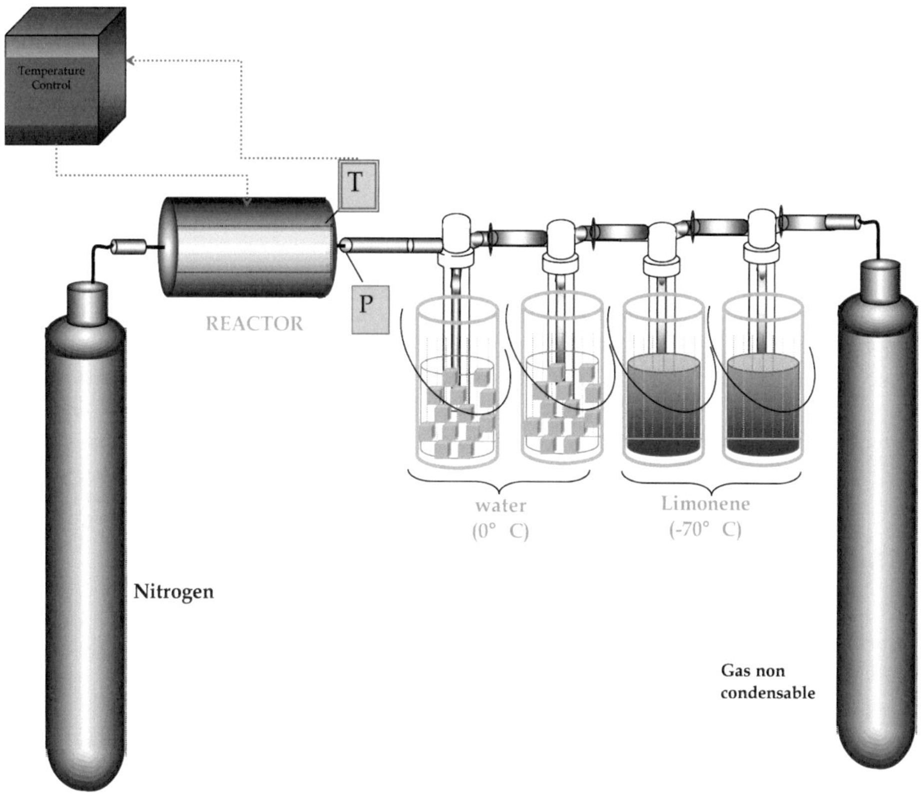

Figure. 1. The schematic diagram of the pyrolysis batch system. T—Thermocouple and P-Pressure transductor.

2.1 Raw material

Two types of biomass were used in the present study. Specifically, pecan nut shell was used as feedstock, obtained from a trading industry of nut located in the city of Torreon, Coahuila in the northeast of México. The orange peel sample was was obtained from a juice processing factory in Monterrey, N.L., México. For easy storage and management, the samples were cut into small pieces of an average area of 2 cm^2 and dried in an air-tunnel at room temperature for the orange peel and in a stove for the case of nut shell.

The elemental analysis of the major components was carried out in a Perkin-Elmer 2400. Moisture was determined by the weight loss at 105 °C for 12 h and is expressed as a weight percentage of the dry matter. The calorific value was obtained in a calorimetric bomb from Parr (model 1341) and it can be expressed in two forms: the gross or higher heating value (HHV) and the net calorific value or lower heating value (LHV). HHV was determined using the calorimeter bomb, through the determination of the temperature difference before and after the occurrence of the sample combustion. LHV is obtained by subtracting the latent heat of vaporization of the water vapor formed by the combustion from the gross or higher heating value.

Ash residue was obtained by thermogravimetry (ASTM D5630 method). Proximate analysis was performed using a thermogravimetric analyzer. The sample was heated under an inert atmosphere at 850 °C and the weight loss during this step is the volatile matter (VM). The gas atmosphere is then switched to air to burn off fixed carbon (FC), while the temperature is reduced to 800 °C. Finally, any residue left after the system is cooled to room temperature and is considered ash.

2.2 Product Characterization

All the collected liquid fractions were characterized by GC/MS using an Agilent Technologies 6890 GC coupled to a 5973 MS. The capillary column was a HP-1, 30 m large, 0.025 mm ID, and helium UHP as the carrier gas. A NIST library in the GC–MS chemstation is used as reference to identify the components of the bio-oil. The last solid residue from the different pyrolysis runs was analysed by elemental analysis. Figure 1 illustrates the schematic diagram of the process, as well as the places where the temperature and pressure were measured and recorded every 30 seconds by means of a data acquisition system.

3. Results of biomass pyrolysis

3.1 Biomass Properties

Biomass is a complex solid material constructed from oxygen-containing organic polymers produced by natural process. The major structural chemical components with high molar masses are carbohydrate polymers and oligomers (65-75%) and lignin (18-35%). The major constituents consist of cellulose (a polymer glucosan), hemicelluloses (which are also called polyose), lignin, organic extractives and inorganic minerals.

The weight percent of cellulose, hemicellulose, and lignin varies in different biomass species. Other biomass compounds are lipids, proteins, simple sugars and water. The pyrolytic chemistry differs sharply between plant carbohydrate polymers from fossil feeds due to the presence of large amounts of oxygen (Mohan et al., 2006). The elemental analysis information of the sweet orange dry peel and pecan nut shell in comparison with other results reported in literature are given in Table 1. The contents of C, H, O and N vary significantly for different types of biomass. The sulphur content in the orange peel is lower than the corresponding to fossil fuels like bituminous coal (4.7 %, wt.) and could be considered as a renewable fuel with lower emission of SOx, which causes pollution and climate change (Sudiro & Bertucco, 2007). Biomass has higher contents of O and H and a lower C content than those reported for fossil fuels.

Biomass	C	H	O	N	S	H/C ratio	O/C ratio	Reference
Almond shell	47.63	5.71	44.48	[a]	[a]	1.44	0.700	Balci et al. (1993)
Coconut shell	47.97	5.88	45.57	0.30	[a]	1.47	0.712	Fagbemi et al. (2001)
Corn cob	43.04	6.32	49.26	1.02	[a]	1.76	0.858	Ren et al. (2009)
Corn Cob	42.90	6.40	49.22	0.60	[a]	1.79	0.860	Yanik et al. (2007)
Cottonseed cake	49.29	5.59	38.67	1.23	[a]	1.36	0.588	Özbay et al. (2001)
Groundnut shell	48.27	5.70	39.40	0.80	[a]	1.42	0.612	Raveendran et al. (1996)
Hazelnut shell	49.94	5.65	42.81	0.27	[a]	1.36	0.643	Balci et al. (1993)
Hazelnut shell	50.08	5.13	41.99	1.38	[a]	1.23	0.629	Demirbas (2006)
Hazelnut shell	50.34	5.84	42.33	0.40	[a]	1.39	0.631	Bonelli et al. (2003)
Orange peel	39.7	6.20	53.0	0.46	0.60	1.87	1.001	Miranda et al. (2009)
Peanut shell	46.59	6.00	43.65	2.06	[a]	1.55	0.703	Bonelli et al. (2003)
Pecan nutshell	47.3	6.40	45.5	0.70	[a]	1.62	0.721	Present work
Pine needles	45.81	5.38	46.11	0.98	[a]	1.41	0.755	Safi et al. (2004)
Rice Straw	45.14	5.85	47.73	0.62	[a]	1.56	0.793	Ren et al. (2009)
Rice Straw	43.68	5.70	39.72	0.97	[a]	1.57	0.682	Xiao et al. (2010)
Sunflower shell	47.40	5.80	41.40	1.40	[a]	1.47	0.655	Demirbas (2006)
Walnut shell	50.58	6.41	41.21	0.39	[a]	1.52	0.611	Onay et al. (2004)
Wheat Straw	48.32	2.54	48.21	0.82	[a]	0.63	0.748	Ren et al. (2009)
Xylan from oat spelts	43.55	5.77	46.33	4.00	0.24	1.59	0.798	Miranda et al. (2009)
Sigmacell	35.45	5.54	57.87	0.82	0.32	1.88	1.224	Miranda et al. (2009)
Kraft lignin	41.06	6.88	50.98	0.65	0.43	2.01	0.931	Miranda et al. (2009)

[a] This data is not available

Table 1. Elemental Analysis of different type of biomass wastes reported in literature.

On the other hand, biomass contains between 36-52 %, wt. carbon while the coal carbon content is about 75-90 %, wt. This means that the heating value of biomass is lower due to the lower energy contained in carbon–oxygen and carbon–hydrogen bonds than those reported for carbon–carbon bonds (Baxter, 1993).

An essential parameter to compare biomaterials or products derived from thermal processes is the elemental composition. The significance of the O:C and H:C ratios of a material on the calorific value can be illustrated using a Van Krevelen diagram, see Figure 2 (Van-Krevelen, 1950). The values for H/C and O/C depend on feedstock, operating conditions, any further treatment methods and water content. Figure 2 shows these ratios for various biomasses and products. For example, these parameters may vary significantly for different biomasses and they include: methanol (H/C, O/C) = (4, 1), methane (4, 0), to various biomass sources (1.2-1.7, 0.6-0.8), pyrolysis oils (1.6, 0.35), pyrolytic carbon (0.13, 0.2) (Miranda et al., 2009), anthracite (0.4, 0.01), lignite (1.14, 0.24) and activated carbon (0.30, 0.04). A typical data for diesel/gasoline is oxygen content close to zero and H/C ratio of 1.5 to 2.

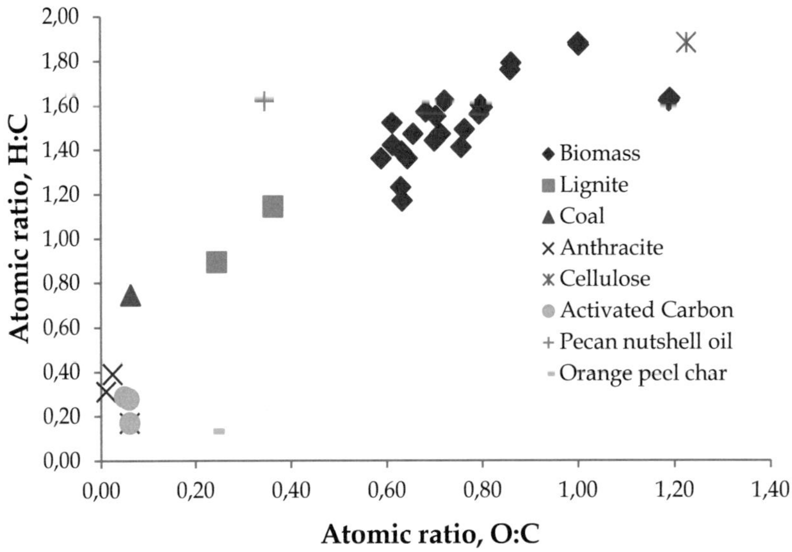

Figure 2. Van-Krevelen diagram of various materials.

Biomass	Volatile %, wt.	Moisture %, wt.	HHV kJ/kg	FC %, wt.	Ash %, wt.	Reference
Orange peel	77.73	9.20	16829	13.07	2.94	Miranda et al. (2009)
Pecan nutshell	60.00	10.28	N.D.	29.72	0.93	Guevara (2009)
P. yezoensis	36.8	9.20	10600	22.10	31.30	Li et al. (2011)
Wheat Straw	76.00	8.10	18910	16.40	7.60	Giuntoli et al. (2009)
Rice husk	62.43	7.16	13620	14.98	15.43	Pütün et al. (2004)
Sugarcane bagasse	84.83	a	20000	13.30	1.89	Das et al. (2004)
Lemon peel	69.84	9.09	17734	17.22	3.85	Heikkinen et al. (2004)
Rice Straw	71.70	a	17100	18.58	9.72	Xiao et al. (2010)
Pine Wood	78.54	6.34	18600	14.66	0.46	Hassan et al. (2009)
Rice Hull	61.00	1.90	a	24.00	13.00	Teng et al. (1998)
Oat Straw	75.90	6.70	17000	0.10	17.30	Ates et al. (20089
Beech	73.62	7.03	a	19.11	0.24	Gómez et al. (2009)
Pine sawdust	80.20	7.90	a	10.80	1.10	Guoxin et al. (2009)
Corn Cob	71.80	8.64	16190	17.50	2.41	Zhang et al. (2009)
White Pine	83.00	2.40	a	14.30	0.30	Lin et al. (2010)
White Oak	85.90	a	8313	13.60	0.50	Gaston (2011)

[a] Data is not available

Table 2. Proximate analysis of different type of biomass.

Table 2 shows the proximate analysis and the calorific value of different materials. This set of analysis gives information on volatile matter (VM), ash and fixed carbon (FC) of a solid biofuels. VM and ash were experimentally quantified, while FC is determined by difference excluding the ash and moisture contents. The volatile content of a solid fuel is that portion released as a gas including moisture by thermogravimetry. VM and FC content provide the measure of chemical energy stored in a solid fuel.

Calorific value is a measure of heating power and is dependent on the composition of the biomass. CV refers to the amount of energy released when a known volume of gas is completely combusted under specified conditions. The significance of the calorific value is that the value provides the total energy content released when the fuel is burning in air. Therefore, CV represents the amount of energy potentially recovered from a given biomass.

3.2 Yield of pyrolysis products

As stated, pyrolysis is the process of the thermal decomposition of organic components in biomass in the absence of oxygen at various temperatures. Biomass can be converted to biochar and bio-oil (carbon rich solid residue and light gases), which can be used to supply the energy requirement of pyrolysis process operations (Bridgwater, 2004; Garcia-Perez et al., 2008a; 2008b). Biomass pyrolysis products are a complex combination of the products from the individual pyrolysis of cellulose, hemicellulose, lignin and extractives; each component has its own kinetic characteristics. In addition, secondary reaction products result from cross-reactions of primary pyrolysis products and reactions between pyrolysis products and the original feedstock molecules (Mohan et al., 2006).

Pyrolysis is one of the most thermally efficient processes to obtain liquid. The material balances of the pyrolysis products of different biomasses are given in Table 3.

At around 700 °C, the weight loss of the orange peel pyrolysis was 78 %, wt., of which 20 %, wt. are light liquid hydrocarbons and 33.90 %, wt. is the heavy fraction, with 24.1 %, wt. as final residue and 22 %, wt. as non-condensable gases. Pyrolysis carried out at 750 °C results in the decrement of the char yield when the pyrolysis temperature increases, while the volatile content increases. From these results, it is evident that an appropriate selection of the heating rate, pyrolysis atmosphere and temperature will lead to more desirable end products. There is a good agreement between thermogravimetric weight loss data previously reported (Guevara, 2009; Miranda et al., 2009) and the data from the fixed bed reactor set-up. Therefore, these results will be helpful for designing and operating a pyrolysis plant of biomass. Bio-oil production converts up to 50-90 %, wt. of biomass energy into the liquid (Huber et al., 2006), which is favorable for fuel handling and transport.

Table 3 shows that yield and composition of pyrolysis products may vary depending on feedstock (Chiaramonti et al., 2007), reactor configurations and pyrolysis conditions (Lou et al., 2004; Bridgwater et al., 2007; Garcia-Perez et al., 2007a, 2007b). Low temperature and long volatiles residence time promote the production of biochar. A high temperature and long residence time increase the cracking of volatiles and, hence gas yield, while a moderate temperature and a short volatiles residence time are optimum for producing bio-oil (Bridgwater et al., 2007). On the other hand, biochar is a good alternative solid fuel for bioenergy production. Higher temperatures lead to lower char yield in all pyrolysis reactions, where the temperature is the main controlling variable of pyrolysis reaction kinetics (Antal & Grønli, 2003).

Biomass	Reactor	Temperature °C	Yield %, wt. solid	Liquid	Gas	Reference
Orange peel	Semi-continuous	700 750	22.0 20.4	53.9 55.3	24.1 24.3	Present work
Pecan nutshell	Semi-continuous	600 700 750	28.0 25.8 20.7	49.2 50.4 54.5	22.8 23.8 22.8	Present work
Corn cob	Tubular	600	24.0	34.0	42.0	Cao et al. (2004)
Rice husk	Fluidized bed	400 450 500 550 600	33.0 32.0 29.0 26.8 25.5	46.5 43.5 37.0 28.5 21.5	6.5 10.0 17.5 25.4 34.5	Williams & Nugranad (2000)
Olive pit	Batch	600	29.0	18.0	53.0	Zabaniotou et al. (2000)
Rice straw	Free-fall	800	84.3	1.0	14.7	Zanzi et al. (2002)
Rice husk	Fluidized bed	420 450 480 510 540	35.0 29.0 24.0 21.0 18.0	53.0 56.0 56.0 33.0 49.0	12.0 15.0 20.0 26.0 33.0	Zheng et al. (2006)
Almond shell	Fixed bed	300 400 500 600 700 800	47.3 30.6 26.0 23.5 21.7 21.5	41.3 53.1 49.3 44.3 36.3 31.0	11.4 16.3 24.7 32.2 42.0 47.5	González et al. (2005)
Rice straw	Fluidized bed	400 412	23.0 32.0	57.0 50.0	20.0 18.0	Lee et al. (2005)
Pecan nutshell	Not available	480 400	17.0 33.0	23.0 44.0	50.0 23.0	Manurung et al. (2009)
Physic nutshell	Fixed bed	500 600 700 800	45.0 42.0 42.0 41.0	30.0 29.0 27.0 26.0	25.0 29.0 31.0 33.0	Sricharoenchaikul et al. (2008)

Table 3. Pyrolysis yields for various biomasses at different conditions.

3.3 Characterization of bio-oils

Bio-oil is clean, cost-effective, CO_2-neutral, and easy to transport and has low sulfur content, making biomass a dominant choice for the replacement of fossil fuels (Nader et al., 2009). Pyrolysis oils are composed of differently sized molecules, which are derived primarily from the de-polymerization and de-fragmentation reactions of the components of the original biomass, mainly cellulose, hemicellulose and lignin (Mohan et al., 2006; Neves et al., 2011). A chromatogram of the bio-oil orange dry peel pyrolysis is shown in Figure 3, where the main peak is located at a retention time of 16.3 min, which is identified as δ-limonene. Table 4 and 5 show the composition of the liquid fractions obtained from the pyrolysis of orange peel and pecan nut shell by GC/MS (see Figures 3 and 4). Nearly, all the components are aromatic compounds. The molecular chains of complex compounds in the orange peel have been broken, generating compounds with a carbon number range of 6–16, see Table 4.

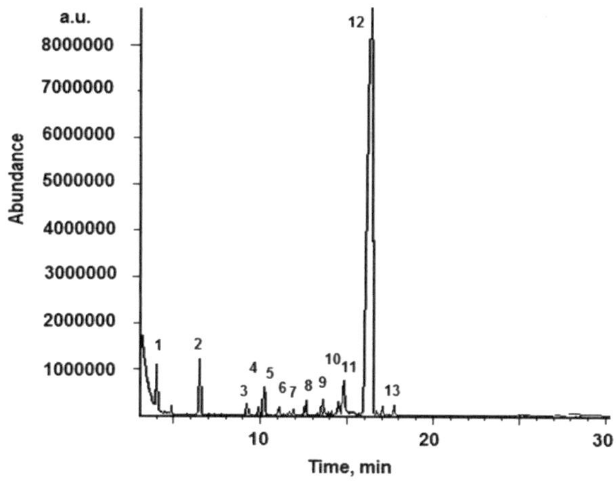

Figure 3. GC–MS spectrum of the bio-oil from dry orange peel pyrolysis bio-oil.

Peak	Retention time, min	Identified Compound
1	4.00	Benzene
2	5.53	Toluene
3	9.43	2-methyl- 2-Hexanol
4	10.01	Ethylbenzene
5	10.25	*p* -Xylene
6	11.18	Styrene
7	11.74	2-Cyclopenten-1-one,2-methyl-
8	12.35	1R-Pinene
9	13.57	Benzene 1-ethyl-3-methyl-
10	14.68	Phenol
11	14.83	β-Pinene
12	16.30	δ-limonene
13	16.96	Phenol, 2-methyl-
14	36.51	n-Hexadecanoic acid

Table 4. Main components of bio-oil from orange peel pyrolysis identified by GC–MS.

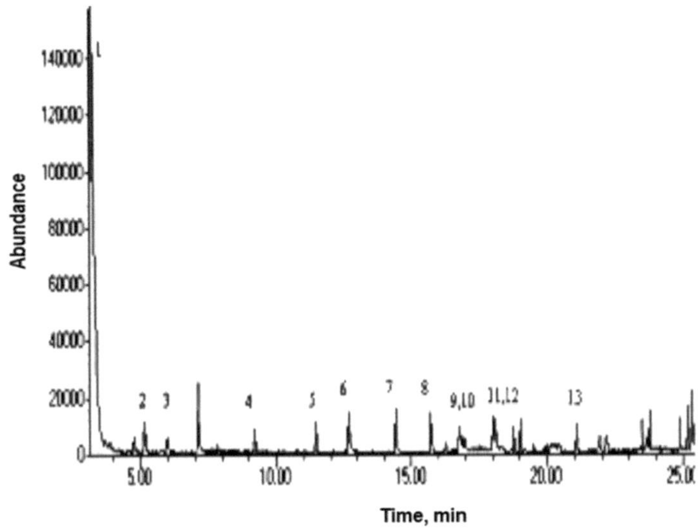

Figure 4. GC–MS spectrum of the bio-oil from dry nut shell pyrolysis bio-oil.

Pyrolysis oil from biomass is a red-brown liquid with pH 3 and 4. Table 7 reports the properties of the pyrolitic oil from both biomasses under study. The oil contains the de-fragmented parts of the oxygenated components of the original biomass structure (mainly cellulose, hemicellulose and lignin). The bio-oil contains oxygen in the range of of 28-40%, wt. oxygen on a dry basis. The bio-oil (i.e., organic phase) has 32 to 42 heating value (MJ/kg) HHV, which is low heating value with respect to fossil fuels. Similar results are reported by literature (Diebold, 2000; Czernik & Bridgwater, 2004; Oasmaa & Meier, 2005; Oasmaa et al., 2005).

Peak	Retention Time, min	Identified compound
1	3.144	Benzene, 1,3-bis(3-phenoxyphenoxy)-
2	5.124	Toluene
3	5.953	2-Pentanone, 3-methylene-
4	9.221	Benzene, (1-methylethyl)-
5	11.487	Limonene
6	12.677	Phenol, 2-methoxy-
7	14.422	Phenol, 2-methoxy-4-methyl-
8	15.732	Phenol, 4-ethyl-2-methoxy-
9	16.264	2-Methoxy-4-vinylphenol
10	16.762	Phenol, 2,6-dimethoxy-
11	18.095	Phenol, 2-methoxy-4-(1-propenyl)-
12	18.799	Phenol, 2,4-bis(1,1-dimethylethyl)-
13	21.088	Phenol, 2,6-dimethoxy-4-(2-propenyl)-

Table 5. Main components of bio-oil from nut shell pyrolysis identified by GC–MS.

Properties of pyrolytic oil	Nut shell	Orange peel
Water content (%-wt.)	30	35
Acidity (pH)	3	4
Elemental composition (%,wt.)		
C	62.40	53.90
H	8.42	6.00
O	28.72	40.0
N	0.30	0.10
Ash content (%, wt.)	0.10	0.10
Heating value (MJ/kg) HHV[a]	32.98	42.70

[a] HHV from organic fraction

Table 6. Properties of pyrolysis bio-oils from nutshell and orange peel.

Pyrolysis liquids are formed by rapidly and simultaneously depolymerizing and fragmenting cellulose, hemicellulose, and lignin with a rapid increase in temperature followed by a condensation system in order to collect all condensable volatiles. If the residence time at high temperature was extended, many products would further react (degrade, cleave, or condensate with other molecules). Bio-oils contain many reactive species, which contribute to unusual attributes (Mohan et al., 2006). Chemically, bio-oil is a complex mixture of components and is difficult to analyze and characterize. Different classes of chemicals are reported in literature such as: water, guaiacols, catecols, syringols, vanillins, furancarboxaldehydes, isoeugenol, pyrones, acetic acid, formic acid, and other carboxylic acids. Bio-oil also contains other major groups of compounds, including hydroxyaldehydes, hydroxyketones, sugars, carboxylic acids, and phenolics. Oligomeric species in bio-oil are derived mainly from lignin, but also from cellulose (Mohan et al., 2006; Oasmaa et al., 2005).

Due to the presence of large amounts of (potentially) highly reactive components, the bio-oil is unstable and tends to separate and forms solids upon storage. Phase separation is promoted by higher temperatures and appears to be faster when the amount of water in the oil is low. Severe polymerization of the oil will result in the formation of char. Distillation also causes undesirable chemical changes leading to the formation of large amounts of solid material. The non- condensable gas products obtained by biomass under study are a mixture of carbon dioxide, carbon monoxide, hydrogen, methane and small amounts of other lighter hydrocarbons, similar to those reported in literature (Bridgwater et al., 1999; Bridgwater & Peacocke, 2000).

4. Conclusion

Pecan nut shell and orange peel are excellent feedstocks for production of energy and value-added products. Biomass residues store a large amount of energy, which can be converted to several forms of usable energy through a number of commercially available processes. Pyrolysis is believed to be the reasonable choice to convert orange peel and pecan nut shell residues to liquid fuels, biocha, and activated carbons. The biomass solid waste in the form of pecan nut shell and orange peel is successfully converted into liquid, char and gas by fixed bed pyrolysis system. The heating value of the pyrolysis oil is found to be 32.98

MJ/kg for pecan nut shell and 42.70 MJ/kg for orange peel bio-oil, which is higher than other biomass-derived pyrolysis oils and also significantly higher than that obtained for the original waste. The maximum liquid yield is found to be 55.3 %, wt. and 54.5 wt% of dry biomass feedstock at the temperature range of 700-750°C and 600-750°C for orange peel and pecan nut shell, respectively. The oil from the biomass under study may be considered as an important important potential potential source of alternative fuel. A lot of research and development process will be necessary in this topic. However, this will occur with greater economic incentive and the climate change regulation will promote research activities in this direction.

5. Acknowledgments

The authors gratefully acknowledge the support of the Chemical Engineering Department of the UANL and CONACYT (México).

6. References

[1] Abdel-Nasser, A. & El-Hendawy (2005). Surface and adsorptive properties of carbon prepared from biomass. *Applied Surface Science,* Vol. 252, No. 2, (October 2005), pp. (287-295), ISSN 0169-4332.

[2] Ahmad, A.L., Loh, M.M. & Aziz, J.A. (2006). Preparation and characterization of activated carbon from oil palm wood and its evaluation on methylene blue adsorption. *Dye and Pigments*, Vol. 75, No. 2, (2007), pp. (263-272), ISSN 0143-7208.

[3] Antal, M.J. & Grønli, M. (2003). The art, science and technology of charcoal production. *Industrial and Engineering Chemistry Research,* Vol. 42, No. 8, (March 2003), pp. (1619-1640), ISSN: 0140-6701.

[4] Ates, F. & Işikdağ, M.A. (2008). Evaluation of the role of the pyrolysis temperature in straw biomass samples and characterization of the oils by GCMS. *Energy & Fuels*, Vol. 22, No. 3, (April 2008), pp. (1936-1943), ISSN 1520-5029.

[5] Balci, S., Dogu, T. & Yucel, H. (1993). Pyrolysis kinetics of lignocellulosic materials. *Industrial &Engineering Chemistry Research*, Vol. 32, No. 11, (November, 1993), pp. (2573-2579), ISSN 08885885.

[6] Baquero, M.C., Giraldo, L., Moreno, J.C., Suárez-García, F., Martínez-Alonso, A. & Tascón, J.M.D. (2003). Activated carbons by pyrolysis of coffee bean husks in presence of phosphoric acid. *Journal of Analytical and Applied Pyrolysis*, Vol. 70, No. 2, (December 2003), pp. (779-784), ISSN 0165-2370.

[7] Baxter, L.L. (1993). Ash deposition during biomass and coal combustion: A mechanistic approach. *Biomass and Bioenergy*, Vol. 4, No. 2, (1993), pp. (85-102), ISSN 0961-9534.

[8] Bonelli, P.R., Cerrella, E.G. & Cukierman, A.L. (2003). Slow pyrolysis of nutshells: Characterization of derived chars and of process kinetics. *Energy Sources*, Vol. 25, No. 8, (Jun 2003), pp. (767-778), ISSN 0090-8312.

[9] Bridgwater, A.V., Carson, P. & Coulson, M. (2007). A comparison of fast and slow pyrolysis liquids from mallee. *International Journal of Global Energy Issues,* Vol. 27, No. 2, pp. (204-216), ISSN 1741-5128.

[10] Bridgwater, A. V. (2004). Biomass Fast Pyrolysis. *Thermal Science*, Vol. 8, No. 2, (November 2004), pp. (21-49), ISSN 0354-9836.

[11] Bridgwater, A.V. & Peacocke, G.V.C. (2000). Fast Pyrolysis Processes for Biomass. *Renewable and Sustainable Energy Reviews*, Vol. 4, No. 1, (March 2000), pp. (1-73), ISSN 1364-0321.

[12] Bridgwater, A.V., Meier, D. & Radlein, D. (1999). An overview of fast pyrolysis of biomass. *Organic Geochemistry*, Volume 30, No. 12, (December 1999), pp. (1479-1493), ISSN 0146-6380.

[13] Budinova, T., Ekinci, E., Yardin, F., Grimm, A., Björnbom, E., Minkova, V. & Goranova, M. (2006). Characterization and application of activated carbon produced by H_3PO_4 and water vapour activation. *Fuel Processing Technology*, Vol. 87, No. 10, (October 2006), pp. (899-905), ISSN 0378-3820.

[14] Cao, Q., Xie, K-C., Bao, W-R. & Shen, S-G. (2004). Pyrolytic behavior of waste corn cob. *Bioresource Technology*, Vol. 94, No. 1, (August 2004), pp. (83-89), ISSN 0960-8524.

[15] Chiaramonti, D., Oasmaa, A. & Solantausta, Y. (2007). Power generation using fast pyrolysis liquids from biomass, *Renewable and Sustainable Energy Reviews*. Vol. 11, No. 6, (August 2007), pp. (1056-1086), ISSN 1364-0321.

[16] Czernik, S. & Bridgwater, A.V. (2004). Overview of Applications of Biomass Fast Pyrolysis Oil. *Energy & Fuels*, Vol. 18, No.2, (February, 2004), pp. (590-598), ISSN 1520-5029.

[17] Das, P., Ganesh, A. & Wangikar, P. (2004). Influence of pretreatment for deashing of sucarcane bagasse on pyolysis products. *Biomass and Bioenergy*, Vol. 27, No. 5, (November 2004), pp. (445-457), ISSN 0961-9534.

[18] Demirbas, A. (2006). Effect of temperature on pyrolysis products from four nut shells. *Journal of Analytical and Applied Pyrolysis*, Vol. 76, No. 1-2, (June 2006), pp. (285-289), ISSN 0165-2370.

[19] Diebold, J. P.(2000). Review of the Chemical and Physical Mechanisms of the Storage Stability of Fast Pyrolysis Bio-Oils. *National Renewable Energy Laboratory*, (January 2000). NREL/SR-570-27613.

[20] Durán-Valle, C.J., Gómez-Corzo, M., Pastor-Villegas, J. & Gómez-Serrano, V. (2005). Study of cherry stone as raw material in preparation of carbonaceous adsorbents. *Journal of Analytical and Applied Pyrolysis*, Vol. 73, No. 1, (March 2005), pp. (59-67), ISSN 0165-2370

[21] Fagbemi, L., Khezami, L. & Capart, R. (2001). Pyrolysis products from different biomasses: application to the thermal cracking of tar. *Applied Energy*, Vol. 69, No. 4, (293-306), pp. (293-306), ISSN 0306-2619.

[22] Garcia-Perez, M., Adams, T.T., Goodrum, J.W., Geller, D.P. & Das, K.C. (2007b). Production and Fuel Properties of Pine Chip Bio-oil/Biodiesel Blends. *Energy & Fuels*, Vol. 21, No. 4, (May 2007), pp. (2363–2372), ISSN 1520-5029.

[23] Garcia-Perez, M., Chaala, A., Pakdel, H., Kretschmer, D. & Roy, D. (2007a). Vacuum pyrolysis of softwood and hardwood biomass: Comparison between product yields and bio-oil properties. *Journal of Analytical and Applied Pyrolysis*, Vol. 78, No. 1, (January 2007), pp. (104-116), ISSN 0165-2370.

[24] Garcia-Perez, M., Wang, S., Shen, J., Rhodes, M., Lee, W.J. & Li, C.Z. (2008a). Effects of Temperature on the Formation of Lignin-Derived Oligomers during the Fast Pyrolysis of Mallee Woody Biomass. *Energy & Fuels*, Vol. 22, No. 3, (March 2008), pp. (2022-2032), ISSN 1520-5029.

[25] Garcia-Perez, M., Wang, X.S., Shen, J., Rhodes, M.J., Tian, F., Lee, W.J., Wu, H. & Li, C.Z. (2008b). Fast Pyrolysis of Oil Mallee Woody Biomass: Effect of Temperature on the Yield and Quality of Pyrolysis Products. *Industrial and Engineering Chemical Research,* Vol. 47, No. 6, (March 2008), pp. (1846-1854), ISSN 1520-5045.

[26] Gaston, K.R., Jarvis, M.W. Pepiot, P., Smith, K.M., Frederick, W.J. & Nimlos, M.R. (2011). Biomass pyrolysis and gasification of varying particle sizes in a fluidized bed reactor. *Energy & fuels,* Vol. 25, No. 8, (August 2011), pp. (3747-3757), ISSN 1520-5029.

[27] Giuntoli, J., Arvelakis, S., Spliethoff, H., de-Jong, W. & Verkooijen, A.H.M. (2009). Quantitative and kinetic thermogravimetric Fourier Transform Infrares (TG-FTIR) study of pyrolysis of agricultural residues. Influence of different pretreatments. *Energy & Fuels,* Vol. 23, No. 11, (August 2009), pp. (5695-5706), ISSN 08870624.

[28] Gómez, C., Velo, E., Barontini, F. & Cozzani, V. (2009). Influence of secondary reactions on the heat of Pyrolysis of biomass. *Industrial and Engineering Chemical Research,* Vol. 48, No. 23, (December 2009), pp. (10222-10233), ISSN 1520-5045.

[29] González, J. F., Ramiro, A., González-García, C. M., Gañan, J., Encinar, J.M, Sabio, E. & Rubiales, J. (2005). Pyrolysis of Almond Shells. Energy Applications of Fractions. *Industrial and Engineering Chemical Research,* Vol. 44, No. 9, (March 2005), pp. (3003-3012), ISSN 0888-5885.

[30] Guevara, K., Miranda, R., Sosa, C. & Rodríguez, M. (2009). Obtención de Bio-Combustibles Mediante Pirólisis De Cáscara De Nuez Pecanera. *Revista Salud Pública y Nutrición,* Vol. 47, No. 1-2010, (Septiembre 2009), Edición especial, ISSN 1870-0160.

[31] Guoxin, H., Hao, H. & Yanhong, L. (2009). Hydrogen-rich gas production from pyrolysis of biomass in an autogenerated steam atmosphere. *Energy & Fuels,* Vol. 23, No. 3, (March 2009), pp. (1748-1753), ISSN 1520-5029.

[32] Hassan, el-B.M., Steele, P.H. & Ingram, L. (2009). Characterization of fast pyrolysis bio-oils produced from pretreated pine wood. *Applied Biochemistry and Biotechnology,* Vol. 154, No. 1, (May 2009), pp. (3-13), ISSN 1559-0291.

[33] Heikkinen, J.M., Hordijk, J.C., de-Jong, W. & Spliethoff, H. (2004). Thermogravimetry as a tool to classify waste component to be used for energy generation. *Journal of Analytical and Applied Pyrolysis,* Vol. 71, No. 2, (June 2004), pp. (883-900), ISSN 0165-2370.

[34] Huber, G.W., Iborra, S. & Corma, A. (2006). Synthesis of Transportation Fuels from Biomass. *Chemistry, Catalysts, and Engineering, Chemical Reviews,* Vol. 106, No. 9, (June 2006), pp. (4044-4098), ISSN 4044-4098.

[35] Lee, K-H., Kang, B-S., Park, Y-K. & Kim, J-S. (2005). Influence of Reaction Temperature, Pretreatment, and a Char Removal System on the Production of Bio-oil from Rice Straw by Fast Pyrolysis, Using a Fluidized Bed. *Energy & Fuels,* Vol. 19, No. 5, (May 2005), pp. (2179-2184), ISSN 0887-0624.

[36] Li, D., Chen, L., Zhang, X., Ye, N. & Xing, F. (2011). Pyrolytic characteristics and kinetic studies of three kinds of red algae. *Biomass and Bioenergy,* Vol. 35, No. 5, (May 2011), pp. (1765-1772), ISSN 0961-9534.

[37] Lin, Y., Zhang, C., Zhang, M. & Zhang, J. (2010). Deoxygenation of bio-oil during pyrolysis of biomass in the presence of CaO in a fluidized-bed reactor. *Energy & fuels,* Vol. 24, No. 11, (September 2010), pp. (5686-5695), ISSN 1520-5029.

[38] Lozano-Castello, D., Alcaniz-Monge, J., Cazorla-Amoros, D., Linares-Solano, A., Zhu, W., Kapteijn, F. & Moulijn, J.A. (2005). Adsorption properties of carbon molecular sieve

prepared from an activated carbon at pitch pyrolysis. *Carbon*, Vol. 43, No. 8, (July 2005), pp. (1643-1651), ISSN 0008-6223.
[39] Lua, A.C., Yang, T. & Guo, J. (2004). Effects of pyrolysis conditions on the properties of activated carbons prepared from pistachio-nut shells. *Journal of Analytical and Applied Pyrolysis*, Vol. 72, No. 2, (November 2004), pp. (279-287), ISSN 0008-6223.
[40] Luo, Z., Wang, S., Liao, Y., Zhou, J., Gu, Y. & Cen, K. (2004). Research on biomass fast pyrolysis for liquid fuel. *Biomass and Bioenergy*, Vol. 26, No. 5, (May 2004), pp. (455-462), ISSN 0961-9534.
[41] Manurung, R., Wever, D.A.Z., Wildschut, J., Venderbosch, R.H., Hidayat, H., Van Dam, J.E.G., Leijenhorst, E.J., Broekhuis, A.A. & Heeres, H.J. 2009. Valorisation of *Jatropha curcas* L. plant parts: Nut shell conversion to fast pyrolysis oil. *Food and Bioproducts Processing*, Vol. 87, No. 3, (September 2009), pp. (187-196), ISSN 0960-3085.
[42] Mehandjiev, D.R., Nickolov, R.N. & Ioncheva, R.B. (1997). Determination of nitrogen structures on activated carbon surfaces by a chemical method. *Fuel*, Vol. 76, No. 5, (April 1997), pp. (381-384), ISSN 0016-2361.
[43] Miranda, R., Bustos-Martínez, D., Sosa-Blanco, C., Gutiérrez-Villareal, M.H. & Rodríguez-Cantú, M.E. (2009). Pyrolysis of sweet orange (*Citrus sinensis*) dry peel. *Journal of Analytical and Applied Pyrolysis*, Vol. 86, No. 2, (November 2009), pp. (245-251), ISSN 0165-2370.
[44] Mohan, D., Pittman, C.U. & Steele, P.H. (2006). Pyrolysis of Wood/Biomass for Bio-oil: A Critical Review. *Energy and Fuels*, Vol. 20, No. 3, (March 2006), pp. (848–889), ISSN 0887-0624 .
[45] Nader, M., Pulikesi, M., Thilakavathi, M. & Renata, R. (2009). Analysis of bio-oil, biogas, and biochar from pressurized pyrolysis of wheat straw using a tubular reactor. *Energy Fuels* Vol 23, No. 1, pp. (2736–2742), ISSN 1520-5029.
[46] Neves, D., Thunman, H., Matos, A., Tarelho, L. & Gómez-Barea, A. (2011). Characterization and prediction of biomass pyrolysis products. *Progress in Energy and Combustion Science*, Vol. 37, No. 5, (September 2011), pp. (611-630), ISSN 0360-1285.
[47] Nickolov, R.N. & Mehandjiev, D.R. (1995). Application of the Simplified equation for micropore size distribution to the study of water vapour adsorption on activated carbon. *Adsorption Science and Technology*, Vol 12, No.3, pp. (203-209), ISSN 0263-6174.
[48] Oasmaa, A. & Meier, D. (2005). Norms and Standards for Fast Pyrolysis Liquids - 1. Round Robin Test. *Journal of Analytical and Applied Pyrolysis*, Vol. 73, No. 2, (June 2005), pp. (323-334), ISSN 0165-2370.
[49] Oasmaa, A.; Peacocke, C., Gust, S., Meier, D. & McLellan, R. (2005). Norms and Standards for Pyrolysis Liquids. End-user Requirements and Specifications. *Energy & Fuels*, Vol. 19, No. 5, (August 2005), pp. (2155-2163), ISSN 0887-0624.
[50] Onay, O., Beis, S.H. & Kockar, O.M. (2004). Pyrolysis of walnut shell in a well-swept fixed-bed reactor. *Energy Sources*, Vol. 26, No. 8, (June 2004), pp. (771-82), ISSN: 0360-5442.
[51] Özbay, N., Pütün, A.E., Uzun, B.B. & Pütün, E. (2001). Biocrude from biomass: pyrolysis of cottonseed cake. *Renewable Energy*. Vol 24, No. 3-4, (November 2001), pp. (615-625), ISSN: 1364-0321.
[52] Prakash, N. & Karunanithi, T. (2008). Kinetic Modeling in Biomass Pyrolysis – A Review. *Journal of Applied Sciences Research*, Vol. 4, No. 12, pp. (1627-1636), ISSN 1819-544X

[53] Pütün, A.E., Apaydin, E. & Pütün, E. (2004). Rice Straw as a bio-oil source via pyrolysis and steam pyrolysis. *Energy*, Vol. 29, No. 12-15, (October-December 2004), pp. (2171-2180), ISSN 0360-5442.

[54] Raveendran, K., Ganesh, A. & Khilar, K.C. (1996). Pyrolysis characteristics of biomass and biomass components. *Fuel*, Vol. 75, No. 8, (June 1996), pp. (987-98), ISSN 0016-2361.

[55] Ren, Q., Zhao, C., Wu, X., Liang, C., Chen, X., Shen, J., Tang, G. & Wang, Z. (2009) Effect of mineral matter on the formation of NOx precursors during biomass pyrolysis. *Journal of Analytical and Applied Pyrolysis*, Vol. 85, No. 1-2, (May 2009), pp. (447-53), ISSN 0008-6223.

[56] Safi, M.J., Mishra, I.M. & Prasad, B. (2004). Global degradation kinetics of pine needles in air. *Thermochimica Acta*, Vol. 412, No.1-2, (March 2004), pp.(155-62), ISSN 0040-6031.

[57] Sircar, S., Golden, T.C. & Rao, M.B. (1996). Activated carbon for gas separation and storage. *Carbon*, Vol. 34, No. 1, pp. (1-12), ISSN 0008-6223.

[58] Sricharoenchaikul, V., Pechyen, C., Aht-ong & Atong, D. Preparation and Characterization of Activated Carbon from the Pyrolysis of Physic Nut (*Jatropha curcas* L.) Waste. *Energy & Fuels*, Vol. 22, No. 1, (September 2007), pp. (31-37), ISSN 0887-0624.

[59] Sudiro, M. & Bertucco, A. (2007). Synthetic Fuels by a Limited CO2 Emission Process Which Uses Both Fossil and Solar Energy. *Energy & Fuels*, Vol. 21, No. 6, pp. (3668-3675), ISSN 08870624.

[60] Teng, H. & Wei, Y.C. (1998). Thermogravimetric studies on the kinetics of rice hull pyrolysis and the influence of water pretreatment. *Industrial and Engineering Chemical Research*, Vol. 37, No. 10, (September 1998), pp. (3806-3811), ISSN 1520-5045.

[61] Van-Krevelen, D.W. (1950). Graphical-statistical method for the study of structure and reaction processes of coal. *Fuel*, Vol. 29, No. 1, pp. (269-84), ISSN 0016-2361.

[62] Williams, P.T & Nugranad, N. (2000). Comparison of products from the pyrolysis and catalytic pyrolysis of rice husks. *Energy*, Vol. 25, No. 6, (June 2000), pp. (493-513), ISSN 0360-5442.

[63] Xiao, R., Chen, X., Wang, F. &Yu, G. (2010). Pyrolysis pretreatment of biomass for entrained-flow gasification. *Applied Energy*, Vol. 87, No. 1, (January 2010), pp. (149-155), ISSN 0306-2619.

[64] Yanik, J., Kornmayer, C., Saglam, M. & Yüksel, M. Fast Pyrolysis of agricultural wastes: characterization of pyrolysis products. (2007). *Fuel processing technology*, Vol. 88, No. 10, (October 2007), pp. (942-947), ISSN 0378-3820.

[65] Zabaniotou, A.A., Roussos, A. & Koroneos, C. (2000). A laboratory study of cotton gin waste pyrolysis. *Journal of Analytical and Applied Pyrolysis*, Vol. 56, No. 1, (September 2000), pp. (47-59), ISSN 0165-2370.

[66] Zanzi, R., Sjöström, K. & Björnbom, E. (2002). Rapid pyrolysis of agricultural residues at high temperature. *Biomass and Bioenergy*, Vol. 23, No. 5, (November 2002), pp. (357-366), ISSN 0961-9534.

[67] Zhang, H., Xiao, R., Wang, D., Zhong, Z., Song, M., Pan, Q. & He, G. (2009). Catalytic fast pyrolysis of biomass in a fluidized bed with fresh and spent fluidized catalytic cracking (FCC) catalysts. *Energy & Fuels*, Vol. 23, No. 12, (December 2009), pp. (6199-6206), ISSN 1520-5029.

[68] Zheng, J-L., Zhu, X-F., Guo, Q-X. & Zhu, Q-S. (2006). Thermal conversion of rice husks and sawdust to liquid fuel. *Waste Management*, Vol. 26, No. 12, (January 2006), pp. (1430-1435), ISSN 0956-053X.